Haifeng Li

Applications of Machine Learning Techniques to Bioinformatics

Haifeng Li

Applications of Machine Learning Techniques to Bioinformatics

VDM Verlag Dr. Müller

Imprint

Bibliographic information by the German National Library: The German National Library lists this publication at the German National Bibliography; detailed bibliographic information is available on the Internet at http://dnb.d-nb.de.

Any brand names and product names mentioned in this book are subject to trademark, brand or patent protection and are trademarks or registered trademarks of their respective holders. The use of brand names, product names, common names, trade names, product descriptions etc. even without a particular marking in this works is in no way to be construed to mean that such names may be regarded as unrestricted in respect of trademark and brand protection legislation and could thus be used by anyone.

Cover image: www.purestockx.com

Publisher:
VDM Verlag Dr. Müller Aktiengesellschaft & Co. KG, Dudweiler Landstr. 125 a, 66123 Saarbrücken, Germany,
Phone +49 681 9100-698, Fax +49 681 9100-988,
Email: info@vdm-verlag.de

Produced in USA and UK by:
Lightning Source Inc., La Vergne, Tennessee, USA
Lightning Source UK Ltd., Milton Keynes, UK
BookSurge LLC, 5341 Dorchester Road, Suite 16, North Charleston, SC 29418, USA

ISBN: 978-3-639-05440-8

To Karl

"Computers are to biology what mathematics is to physics."

— Harold Morowitz

"The goal is to understand ground truth."

— Richard M. Karp

"Our horizon is never quite at our elbows."

— Henry David Thoreau

"Nothing is more practical than a good theory."

— Vladimir N. Vapnik

Contents

Preface   xi

1 Introduction   1

2 Prediction of Translation Initiation Sites in Eukaryotic
   mRNAs   9

   2.1 Introduction 9

   2.2 Support Vector Machines and Edit Kernel 12

        2.2.1 Support Vector Machines 13

        2.2.2 An Edit Kernel 15

   2.3 Redundancy of Codons and More General Edit Costs . . 20

        2.3.1 Edit Costs Based on Mutation Probabilities . . . 21

   2.4 Experiments 23

        2.4.1 Data 25

2.4.2   Experimental Results   26

2.4.3   Computational Issues   30

2.4.4   Discussion   32

2.5   An Online Program: TISHunter   33

2.6   Prediction on the Human mRNAs   34

2.7   Conclusion   36

3   Accurate Cancer Diagnosis with Gene Expression Profiling   37

3.1   Introduction   37

3.2   Linear Discriminant Analysis   42

3.2.1   The Small Sample Size Problem   43

3.2.2   Previous Work   44

3.3   Generalized Linear Discriminant Analysis   46

3.3.1   Method   47

3.3.2   A Fast Algorithm   52

3.3.3   Discussion   54

3.4   Experiments   56

3.4.1   Data   56

vi

3.4.2   Results   57

3.5   Conclusion   63

4   Gene Expression Analysis with Minimum Entropy Clustering   65

4.1   Introduction   65

4.2   Related Work   68

4.3   Minimum Entropy Clustering Criterion   71

4.4   The Clustering Algorithms   76

     4.4.1   Estimation of *a Posteriori* Probability   76

     4.4.2   An Iterative Algorithm   80

4.5   Experiments   84

     4.5.1   Synthetic Data   85

     4.5.2   Real Gene Expression Data   88

4.6   Conclusion   93

5   A General Framework for Biclustering Gene Expression Data   95

5.1   Introduction   95

5.2   Kolmogorov Complexity   98

vii

5.3   Methods 100

    5.3.1   Clustering and Checkerboard Type Biclustering . 104

    5.3.2   Approximating Kolmogorov complexity 104

5.4   Experiments 109

5.5   Conclusion 117

6 Systematic Discovery of Functional Modules and Context-Specific Functional Annotation of Human Genome   119

6.1   Introduction 119

6.2   Materials and Methods 122

    6.2.1   Microarray Data 122

    6.2.2   Gene Ontology Function Categories 122

    6.2.3   Graph Construction 123

    6.2.4   Mining Recurrent Network Modules 123

    6.2.5   Network Topology Score for Each Function Category 125

    6.2.6   Assessment of Function-Assignments with Random Forest 126

6.3   Results 127

    6.3.1   Systematic Identification of Functional Modules in Human Genome 127

6.3.2   Enrichment of Protein-Protein Interaction in Network Modules 131

6.3.3   Function Prediction 132

6.3.4   Context-Specific Function Annotation 133

6.3.5   Discovery of Uncharacterized Cellular Systems . . 137

6.4   Conclusion and Discussion 138

7 A Quantile Method for Sizing Optical Maps   141

7.1   Introduction 141

7.2   Method 143

7.3   Method Validation 153

7.4   Discussion 157

8 The Regularized EM Algorithm   161

8.1   Introduction 161

8.2   The Regularized EM Algorithm 163

8.3   Finite Mixture Model 167

8.4   Demonstration 171

8.5   Conclusion 176

ix

9 Conclusions   **177**

Appendices   **181**

A Matrix Algebra   **181**

    A.1 Notation 181

    A.2 Eigenvalues and Eigenvectors 182

    A.3 Matrix Derivatives 183

    A.4 Generalized Inverse of Matrices 184

B Derivations of Some Equations   **187**

    B.1 Derivation of Equation (8.19) 187

    B.2 Derivation of Equation (8.23) 189

Bibliography   **191**

Preface

The completion of the sequencing of the human genome was heralded the dawn of a new era in biology and medicine. Besides, advances in high-throughput experimental technologies enable us to observe various aspects of biological processes. For example, DNA microarrays can be used to measure changes in expression levels or to detect of SNPs (single nucleotide polymorphism) at the level of the whole genome. ChIP-on-chip (also known as ChIP-chip) allows the identification of binding sites of DNA-binding proteins in a very efficient and scalable way. Optical mapping, a high-throughput single-molecule system, can construct physical maps spanning entire genomes. Such a global and comprehensive genomic view also changes the landscape of biological and biomedical research. However, the huge amount of genomic data are of limited value if we cannot effectively use them to understand the complicated biological systems and processes.

This book is devoted to developing effective and efficient machine learning techniques for analyzing the huge amount of genetic data. Mathematically, learning means to fit a multivariate function to a given number of samples. Critically, the fitting should be predictive. After training a model on the gene expression profiling of some tumor and normal tissue samples, for instance, we hope that it can accurately determine if a new tissue sample is tumor or normal. More importantly, learning may also help us to discover the underlying biological way by fitting experimental biology data. For example, we may computationally determine the genetic markers related with cancer. In practice, learning techniques also have to be efficient so that we can deal with the flood of genetic data.

Following the above principles, we have proposed several learning methods for sequence annotation, microarray gene expression analysis, and optical mapping, including predicting translation initiation sites in eukaryotic mRNAs with support vector machines and edit kernels, accurate and robust cancer classification with gene expression profiling, minimum entropy clustering method, a general framework for biclustering gene expression data, an integrative approach for microarray gene expression data analysis, and a quantile method for sizing optical maps. Because the EM algorithm has been widely applied in computational biology and bioinformatics, we also develop a regularized EM algorithm that can effectively reduce the uncertainty of missing data.

This book is mainly based on my research work during Ph.D. and postdoc study in the University of California, Riverside and the University of Southern California, respectively. I would like to express sincere gratitude to my Ph.D. advisor, Dr. Tao Jiang, and postdoc advisors, Dr. Michael S. Waterman, Dr. Xianghong Jasmine Zhou and Dr. Lei M. Li, for their invaluable inspiration and guidance. Discussions with them were always very delightful, and I gained considerable knowledge and insight into problems and techniques in bioinformatics, statistical and algorithms areas. Many other professors also helped and encouraged me in various ways during my study and research. I am especially grateful to Profs. Stefano Lonardi, Dimitrios Gunopulos, Marek Chrobak, Fengzhu Sun, and Haiyan Huang for all that they have taught me. I would like to thank the whole Algorithms and Computational Biology Laboratory at the University of California, Riverside and the Center of Excellence in Genomic Science at the University of Southern California, with whom I had the pleasure of carrying out research in an excellent environment. I am particularly thankful to my friends Qi Fu, Xin Chen, Zheng Liu, Yu Huang, Anton Valouev, Haiyan Hu, Xifeng Yan, Li Jia, Andres Figueroa, Jing Li, Zheng Fu, Lan Liu, Jie Zheng, Chuhu Yang, and Qiaofeng Yang. Finally, I am at a loss to find suitable words that express my deep appreciation for the significant contributions of my wife and my parents for their support.

Chapter 1

Introduction

About one hundred and forty years after Mendel's discovery of the law of heredity and about fifty years after Watson and Crick's discovery of the double-helical structure of DNA [Watson and Crick, 1953], scientists now have in hand the complete DNA sequences of genomes for many organisms - from microbes to plants to humans. Genomics has become an very important and extremely active field of biology and biomedicine. Many people believe that the genomic era is a reality now and this century will be "the" century of biology and genomics.

Based on the great accomplishments of Human Genome Project, researchers now aim at many bold and ambitious research targets, which can be formulated into three major themes – genomics to biology (elucidating the structure and function of genomes), genomics to health (translating genome-based knowledge into health benefits), and genomics to society (promoting the use of genomics to maximize benefits and minimize harms) [Collins et al., 2003]. To achieve these goals, computational methods cannot be avoided as well as resources, technology development, etc. In fact, computational methods have become intrinsic to modern biological research. The importance of computational biology lies in not only the huge amount of genetic data due to large-scale data generation technology, but also the extreme complexity of biology systems. It is believed that all future biological research will integrate both computational and experimental components.

There are a lot of challenging and important problems in bioinformatics and computational biology, for example the identification of different features in a DNA sequence, the analysis of gene expression and regulation, the elucidation of protein structure and protein-protein interactions, the determination of the relationship between genotype and phenotype, and the identification of the patterns of genetic variation in populations and the processes that produced those patterns [Collins et al., 2003]. In this book, I will focus on employing machine learning techniques to solve some problems in the identification of gene structure and gene expression analysis. In particular, I will work on the prediction of translation initiation sites in eukaryotic mRNAs, cancer classification with gene expression profiling, clustering and biclustering gene expression data, and the regularized EM algorithm. Although most of proposed methods were initially designed to solve some biological problems, they are actually very general machine learning methods and may find applications in other areas, such as image recognition, web document classification, density estimation, etc. In the rest of the chapter, I will briefly introduce the methods and results.

One of the main goals of the Human Genome Project is to provide a complete list of annotated genes to serve as a "periodic table" for biomedical research. The identification of coding regions in uncharacterized eukaryotic DNA sequences is a central problem in gene prediction. Designing accurate and efficient gene finding algorithms is currently a very active and important research topic in computational biology. My work in this area mainly focuses on the prediction of translation initiation sites (TISs) in eukaryotic mRNAs. Although algorithms for finding the internal coding exons of a gene have reached a high degree of sophistication and accuracy, finding translation initiation sites that encode the start of protein translation, still remains a challenge. In Chapter 2, we present a class of new sequence-similarity kernels, called the edit kernels, for use with support vector machines (SVMs) in a discriminative approach to recognize TISs with very high accuracy. The edit kernels are simple and have significant biological and probabilistic interpretations. Although the edit kernels are not positive definite, it is easy to make the kernel matrix positive definite by adjusting the parameters. We also

2

convert the region of an input mRNA sequence downstream to a putative TIS into an amino acid sequence before applying SVMs to avoid the high redundancy in the genetic code. The algorithm has been implemented and tested on previously published data. Our experimental results on real mRNA data show that both ideas improve the prediction accuracy greatly and our method performs significantly better than those based on neural networks and SVMs with polynomial kernels or Salzberg kernel.

Although the high-quality, comprehensive sequences of the human genome have been available [International Human Genome Sequencing Consortium, 2001, Venter et al., 2001], the challenge still remains to understand the procedure from genome to life. For example, what are the gene regulatory networks that govern which genes are expressed in a cell at any given time, how much products is made from each one, and the cell's responses to diverse environmental cues and intracellular signals. Microarray technology provides us a great opportunity to reveal these biological secrets by simultaneously observing the expression levels of many thousands of genes. Such a global view of thousands of functional genes changes the landscape of biological and biomedical research. To elucidate the patterns hidden in the gene expression data, we develop several methods including accurate cancer classification, clustering, and biclustering.

Robust and accurate cancer classification is critical in cancer treatment. Gene expression profiling is expected to enable us to diagnose tumors precisely and systematically. However, the classification task in this context is very challenging because of the curse of dimensionality and the small sample size problem. In Chapter 3, we propose novel methods to solve these two problems. Our methods are able to map gene expression data into a very low dimensional space and thus meets the recommended samples to features per class ratio. As a result, our methods can be used to classify new samples robustly with low and trustable (estimated) error rates. The method is based on linear discriminant analysis (LDA). However, the conventional LDA requires that the within-class scatter matrix $\mathbf{S}_w$ be nonsingular. Unfortunately, $\mathbf{S}_w$ is always singular in the case of cancer classification due to the small sample size problem. To overcome this problem, we develop the generalized linear discriminant analysis (GLDA). Our method is mathematically well-founded and coin-

3

cide with the conventional LDA when $\mathbf{S}_w$ is nonsingular. Different from the conventional LDA, our method does not assume the nonsingularity of $\mathbf{S}_w$, and thus naturally solves the small sample size problem. To accommodate the high dimensionality of scatter matrices, a fast algorithm is also developed. Our extensive experiments on seven public cancer datasets show that the methods performs well. Especially on some difficult instances that have very small samples to genes per class ratios, our methods achieve the much higher accuracies than the widely used classification methods such as support vector machines, random forests, etc.

A preliminary and common methodology for analyzing gene expression data is the clustering technique. Although many clustering methods have been proposed, most of them cannot either easily determine the number of clusters, or find non-convex clusters, or handle outliers. Although many sophisticated methods have been proposed to address each of aforementioned challenges, it results in a long learning curve for biologists to study many methods on model selection and outlier detection besides clustering. Besides, many parameters are usually involved in the algorithms, which also requires abundant experience for biologists to choose good ones. As a result, biologists often choose simple methods, such as k-means and hierarchical clustering, to analyze gene expression data. Unfortunately, it is well known that these methods are not effective in many situations and may not fit data well. In Chapter 4, we present a simple clustering algorithm to address all above challenges with only one parameter. The input parameter, bandwidth of Parzen window, controls the resolution of clustering. Since most genes are multi-functional and the functions can be roughly organized in to hierarchical structure, such a multi-resolution property is very useful for gene expression analysis. Our method tries to locally minimize the conditional entropy of clusters given observations, which is a very meaningful criterion for clustering as indicated by Fano's inequality and the probability error of the nearest neighbor method. We develop an efficient iterative algorithm to optimize our clustering criterion with a nonparametric approach to estimate *a posteriori* probabilities. The experimental results on various datasets show that the clustering algorithm performs well.

Recently, gene expression data with many heterogeneous conditions

4

appeared. The common clustering methods cannot be applied on these datasets because the assumption that related genes co-express in all conditions is too restricted. The current biclustering methods try to find some type of biclusters but no one can discover all types of patterns in the data. Furthermore, researchers have to design new algorithms in order to find new types of biclusters/patterns that interest biologists. In chapter 5, we propose a novel approach for biclustering that in general can be used to discover all computable patterns in gene expression data. The method is based on the theory of Kolmogorov complexity. More precisely, we use Kolmogorov complexity to measure the randomness of submatrices as the merit of biclusters because randomness naturally consists in a lack of regularity, which is a common property of all types of patterns. Based on algorithmic probability measure, we develop a Markov chain Monte Carlo (MCMC) algorithm to search for biclusters. Our method can be also easily extended to solve the problems of conventional clustering and checkerboard type biclustering. Preliminary experiments on simulated as well as real data show that our approach is very versatile and promising.

Previously, most microarray gene expression analysis methods focus on single dataset. On the other hand, the rapid accumulation of microarray datasets provides unique opportunities to perform systematic functional characterization of the human genome. In chapter 6, we designed a graph-based approach to integrate cross-platform microarray data, and extract recurrent expression patterns. A series of microarray data sets can be modeled as a series of co-expression networks, in which we search for frequently occurring network patterns. The integrative approach provides three major advantages over the commonly used microarray analysis methods: (1) enhance signal to noise separation (2) identify functionally related genes without co-expression, and (3) provide a way to predict gene functions in a context-specific way. We integrate 65 human microarray data sets, comprising 1105 experiments and over 11 million expression measurements. We develop a data mining procedure based on frequent itemset mining and biclustering to systematically discover network patterns that recur in at least 5 datasets. This resulted in 143,401 potential functional modules. Subsequently, we design a network topology statistic based on graph random walk that effectively captures characteristics of a gene's local functional environment. Function annotations based on this statistic are then subject to the assessment us-ing the random forest method, combining six other attributes of the network modules. We assign 1126 functions to 895 genes, 779 known and 116 unknown, with a validation accuracy of 70%. Among our assignments, 20% genes are assigned with multiple functions based on different network environments.

Optical mapping is an integrated system for the analysis of single DNA molecules. It constructs restriction maps (noted as "optical map") from individual DNA molecules presented on surfaces after they are imaged by fluorescence microscopy. Because restriction digestion and fluorochrome staining are performed after molecules are mounted, resulting restriction fragments retain their order. Maps of fragment sizes and order are constructed by image processing techniques employing integrated fluorescence intensity measurements. Such analysis, in place of molecular length measurements, obviates need for uniformly elongated molecules, but requires samples containing small fluorescent reference molecules for accurate sizing. Although robust in practice, elimination of internal reference molecules would reduce errors and extend single molecule analysis to other platforms. In Chapter 7, we introduce a new approach that does not use reference molecules for direct estimation of restriction fragment sizes, by the exploitation of the quantiles associated with their expected distribution. We show that this approach is comparable to the current reference-based method as evaluated by map alignment techniques in terms of the rate of placement of optical maps to published sequence.

The EM algorithm has been widely applied to various problems in computational biology and bioinformatics. The EM algorithm heavily relies on the interpretation of observations as incomplete data. However, the original EM algorithm does not have any control on the uncertainty of missing data. To address this issue, we develop a regularized EM algorithm that can effectively reduce the uncertainty of missing data in Chapter 8. The regularized EM algorithm penalizes the likelihood with the mutual information between the missing data and the incomplete data (or the conditional entropy of the missing data given the observations). The proposed method maintains the advantage of the conventional EM algorithm, such as reliable global convergence, low cost per iteration, economy of storage, and ease of programming. We also apply the regularized EM algorithm to fit the finite mixture model. Our theoretical analysis and experiments show that the new method can efficiently fit the models and effectively simplify over-complicated models.

Chapter 2

Prediction of Translation Initiation Sites in Eukaryotic mRNAs

2.1 Introduction

One of the main goals of the Human Genome Project is to provide a complete list of annotated genes to serve as a "periodic table" for biomedical research. The identification of coding regions in uncharacterized eukaryotic DNA sequences is a central problem in gene prediction. Many algorithms and systems have been developed to automatically find the components of a gene, which include *translation initiation sites* (TISs), exon-intron splice sites, promoters, poly-adenylation signals, and CpG islands.

Although the algorithms (*e.g.* GENSCAN [Burge and Karlin, 1997]) for finding the internal coding exons of a gene have reached a high degree of sophistication and accuracy, finding translation initiation sites that encode the start of protein translation, still remains a challenge.

9

The codon ATG is the most commonly used start codon. [1] Usually, the initiation of translation starts at the first ATG codon in an mRNA. However, sometimes a downstream ATG is selected due to leaky scanning, reinitiation, and internal initiation of translation (this happens only for some viral mRNAs), *etc.* [Kozak, 1996, Kozak, 1999]. According to [Kozak, 1989, Yoon and Donahue, 1992], downstream ATGs are used as start codons in less than 10% of investigated eukaryotic mRNAs. So, it seems that we could easily obtain an accuracy of more than 90% in the prediction of TISs by simply selecting the first ATG, given complete and error-free mRNA sequences. However, it has been reported that, in the GenBank nucleotide data that are annotated as being equivalent to mature mRNAs, almost 40% of the sequences contain upstream ATGs [Pedersen and Nielsen, 1997]. This problem is enhanced when using unannotated genomic data and when analyzing expressed sequence tags (ESTs), which are single-pass partial sequences derived from cDNAs and are usually error-prone. All these problems make it very difficult to predict TISs accurately.

The recognition of TISs has been extensively studied by using biological approaches, machine learning, and statistical models. In 1987, Kozak found that an ATG codon in a very weak context is not likely to be the start site of translation [Kozak, 1987b]. The optimal context for initiation of translation in vertebrate mRNA is GCCACCatgG. Within this *consensus* motif, nucleotides in two highly conserved positions exert the strongest effect: a G residue following the ATG codon (position +4) and a purine, preferably A, three nucleotides upstream (position −3). However, such a consensus alone is not sufficient to identify the ATG initiator codon [Kozak, 1996, Kozak, 1999]. For example, after an 80S ribosome translates the first *small* open reading frame and reaches a stop codon, the 40S subunit may hold on to the mRNA, resume scanning, and reinitiate at a downstream ATG codon. This procedure is called *reinitiation* [Kozak, 1999]. To predict TISs, Kozak developed a weight matrix from an extended collection of data [Kozak, 1987a].

Statistical methods have also been developed to predict TISs. For example, Salzberg developed a *positional conditional probability matrix* that takes into account the dependency between adjacent bases [Salzberg,

---

[1]In some special cases, the codon GTG is used.

10

1997]. Agarwal and Bafna developed the so called *generalized second-order profiles* that consider dependencies between non-adjacent bases [Agarwal and Bafna, 1998]. However, both methods suffer from high rates of false positives.

Since 1997, the machine learning approach has been applied to find TISs. With a neural network, Pedersen and Nielsen achieved a 84.6% accuracy on a collection of 3312 vertebrate sequences [Pedersen and Nielsen, 1997]. Salamov *et al.* used six characteristics to analyze the area around a putative start codon and employed linear discriminant analysis for the final scoring [Salamov et al., 1998]. Zien *et al.* used support vector machines (SVMs) to predict TISs and achieved an 88.6% accuracy on Pedersen and Nielsen's data [Zien et al., 2000]. Recently, Hatzigeorgiou achieved a 94% accuracy on 475 cDNA sequences [Hatzigeorgiou, 2002]. Her system includes two modules (both based on neural networks), one sensitive to the conserved motif and the other sensitive to the coding/noncoding potential around the start codon. The program linearly searches the coding ORF and stops once the combination of the two modules predicts a positive score. We observe that it is meaningless to compare the performance between Hatzigeorgiou's approach and Pedersen and Nielsen's approach here since they were tested on the different data. Finding TISs was also addressed indirectly in [Davuluri et al., 2001] in terms of finding the first exon of a gene contained in a genomic sequence. In [Davuluri et al., 2001], Davuluri *et al.* developed the program FirstEF based on a decision tree consisting of quadratic discriminant functions. Besides TISs, FirstEF can also recognize CpG islands, promoter regions and the first splice-donor sites. Using different models to predict CpG-related and non-CpG-related first exons, FirstEF could predict 86% of the first exons.

In this chapter, we present a new algorithm to recognize TISs with very high accuracy. Our algorithm contains two major ideas. First, we introduce a class of new sequence-similarity kernels, called the *edit kernels*, for use with support vector machines (SVMs) in a discriminative approach to predict TISs. Our kernels are based on the *string edit distance* and have natural biological and probabilistic interpretations. Second, we treat the upstream and the downstream regions of a putative TIS in different ways. More precisely, we convert the downstream

11

region of a putative TIS into an amino acid sequence before applying SVMs to avoid the high redundancy in the genetic code. The similarity between amino acids is also considered. The algorithm has been implemented with several variants of the edit kernel and tested on Pedersen and Nielsen's data set (as well as some smaller data sets derived from this data). The experiments demonstrate that our algorithm can achieve an accuracy of 99.90% with 99.92% sensitivity and 99.82% specificity, which are significantly better than those of the previous algorithms.

The rest of this chapter is organized as follows. Section 2.2 gives a brief review of SVMs and introduces the basic edit kernel using string edit distance. Section 2.3 extends the edit kernel by considering the redundancy in the genetic code and similarity between amino acids, and presents two more sophisticated edit kernels. In Section 2.4, we describe some experimental results on Pedersen and Nielsen's data and small data sets derived from the data. The performance of the edit kernels, the impact of the choice of the edit cost matrix on the performance and efficiency of the SVMs, and some computational issues are discussed in this section. We also give some intuitive explanation on why the SVMs with the above edit kernels work so well. In Section 2.5, we introduce a publicly accessible online program, called *TISHunter*, for predicting TISs based on our algorithm. Section 2.6 describes the prediction results on mRNAs from the human genome. Section 2.7 concludes the chapter with some directions of further research.

2.2 Support Vector Machines and Edit Kernel

Many methods have been proposed for classification problems in bioinformatics. Roughly, these methods follow two approaches, the *generative* approach and the *discriminative* approach. The generative approach, *e.g.* hidden Markov models, builds a model for the target pattern and then evaluates each candidate sequence to see how well it fits the model. If the fitting score is above some threshold, then the candidate is classified into the pattern. The discriminative approach, *e.g.* neural networks, tries to learn some discriminant function from some samples that have been labelled as positive or negative. After learning, the discriminant functions are employed to decide whether a new sample is positive or not. In this chapter, we follow the discriminative approach to recognize TISs. In particular, we will employ SVMs with a new class of sequence-similarity kernels, the edit kernels. In what follows, we first give a brief review of SVMs. Then we introduce an edit kernel and discuss its biological and statistical meaning.

2.2.1 Support Vector Machines

Given a set of independent and identically distributed samples (*i.e.* the training data) in the form pairs of patterns x_i and labels y_i,

$$(x_1, y_1), \ldots, (x_\ell, y_\ell) \in \mathcal{X} \times \{\pm 1\}$$

we want to learn a functional dependency $y = f(x; \alpha)$ between x_i and y_i, where α is a parameter from the set Λ. We hope that $f(x; \alpha)$ could make the smallest number of expected errors on the unseen samples drawn from the same distribution.

For a linearly separable data, SVM is the optimal hyperplane $y = \text{sign}(\langle w, x \rangle + b)$ that maximizes the *margin* $1/\|w\|^2$ between the classes, which is the minimum distance from positive/negative samples to the separation hyperplane [Vapnik, 1995, Vapnik, 1998]. The reason to maximize the margin is that hyperplanes with a larger margin have a smaller capacity (actually a smaller upper bound on the VC-dimension) [Vapnik, 1995, Vapnik, 1998]. In this way, the overfitting problem could be avoided. The optimal hyperplane can be found by maximizing

$$\mathcal{L} = \sum_{i=1}^{\ell} \alpha_i - \frac{1}{2} \sum_{i,j=1}^{\ell} \alpha_i \alpha_j y_i y_j \langle x_i, x_j \rangle \tag{2.1}$$

subject to

$$0 \leq \alpha_i \leq C, \quad i = 1, \ldots, \ell \tag{2.2}$$

$$\sum_{i=1}^{\ell} \alpha_i y_i = 0 \tag{2.3}$$

13

where α_i's are the Lagrange multipliers and C is a positive constant. Solving this quadratic programming problem, we get w, b, and thus the optimal hyperplane

$$y = \text{sign} \left(\sum_i \alpha_i y_i \langle x, x_i \rangle - b \right) \qquad (2.4)$$

According to the well-known Karush-Kuhn-Tucker (KKT) necessary conditions [Bertsekas, 1999], the solution w and b to the above quadratic programming problem must satisfy

$$\alpha_i(y_i(\langle w, x_i \rangle + b) - 1) = 0 \qquad (2.5)$$

It means that only the points x_i on the hyperplanes $\langle w, x_i \rangle + b = \pm 1$ have nonzero Lagrange multipliers α_i. Thus, only these points x_i make effects in the optimal hyperplane. Such points are called *support vectors*.

Usually, the input data is not linearly separable. In this case, the input data is first mapped into a high-dimensional feature space $\mathcal{F}$ via a nonlinear function $\Phi(\cdot) : \mathcal{X} \to \mathcal{F}$. SVMs in this high-dimensional space $\mathcal{F}$ can be applied since the data may be linearly separable after the mapping. However, $\mathcal{F}$ could have an arbitrarily large, possibly infinite, dimensionality, which makes it impossible to map points into $\mathcal{F}$ directly. To overcome this obstacle, SVMs can perform the mapping Φ *implicitly*. This is possible because all information that we need supply to an SVM are the dot products $\langle \Phi(x_i), \Phi(x_j) \rangle$ in the feature space $\mathcal{F}$, which can be computed through a *positive definite kernel* $k(\cdot, \cdot)$ in the input data space [Aizerman et al., 1964, Aronszajn, 1950]:

$$k(x_i, x_j) = \langle \Phi(x_i), \Phi(x_j) \rangle \qquad (2.6)$$

The positive definite kernel (also known as Mercer kernel) is formally defined as follows [Berg et al., 1984]:

Definition 2.1 *Let $\mathcal{X}$ be a nonempty set. A function $k(\cdot, \cdot) : \mathcal{X} \times \mathcal{X} \to \mathbb{R}$ is called a positive definite kernel [2] if $k(\cdot, \cdot)$ is symmetric (i.e. $k(x, y) =$*

---

[2]The definition can be extended to the more general case of complex-valued kernels. In this chapter, we will only need consider real-valued kernels.

$k(y, x)$ for all $x, y \in \mathcal{X}$) and

$$\sum_{i=1}^{n} \sum_{j=1}^{n} c_i c_j k(x_i, x_j) \geq 0 \qquad (2.7)$$

for all $n \in \mathbb{N}$, $x_1, \ldots, x_n \in \mathcal{X}$ and $c_1, \ldots, c_n \in \mathbb{R}$

For example, polynomial kernel $\langle x, y \rangle^d$ and Gaussian kernel $e^{-\gamma \|x-y\|^2}$ are two well-known positive definite kernels. The matrix $K_{ij} = k(x_i, x_j)$ is called the kernel matrix, which is just the Gram matrix of dot products $K_{ij} = \langle \Phi(x_i), \Phi(x_j) \rangle$ in feature space $\mathcal{F}$.

In practice, the kernel value $k(x, y)$ can also be interpreted as the pairwise similarity between x and y. For example, if x and y are unit-length vectors, the simple dot product $\langle x, y \rangle$ computes the cosine of the angle between x and y. Clearly, two "similar" vectors have a small angle between them and thus a large cosine. Therefore, we can think of the kernel value $k(x, y)$ as a measure of the similarity between x and y in the feature space $\mathcal{F}$. More generally, the similarity represented by $k(x, y)$ could be interpreted in the sense of specific applications and does not necessarily follow the above geometrical interpretation.

2.2.2   An Edit Kernel

Before introducing our edit kernel, let us review the approach of Zien et al. who used SVMs with a polynomial kernel based on Hamming similarity to recognize TISs [Zien et al., 2000]. Zien et al. used a sparse-encoding scheme to represent nucleotides: each nucleotide is encoded by five bits, exactly one of which is set. The position of the set bit indicates whether the nucleotide is A, C, G, T, or N (for unknown). Using this encoding scheme, Zien et al. introduced an SVM with the polynomial kernel

$$k(x, y) = \langle x, y \rangle^d$$

Note that, on a sparsely encoded data, the dot product $\langle x, y \rangle$ counts exactly the number of nucleotides that coincide in two sequences represented by x and y, i.e. the complement of Hamming distance between

x and y. In order to handle local correlations of sequences, Zien *et al.* also introduced two variations of polynomial kernel, called the locality-improved kernel and Salzberg kernel. The locality-improved kernel uses a small sliding window to scan the input sequence and counts matching nucleotides in every window. All these counts are raised to the power of d_1 and then are added up. At last, the sum is taken to the power of d_2. Here, d_1 and d_2 are user-specified parameters. Salzberg kernel is similar to the locality-improved kernel. However, instead of the original nucleotide sequences, Salzberg kernel is applied on the sequences of log odds scores $s_p(x)$, which is defined as

$$s_p(x) = \log \frac{P(x_p \text{ at pos. } p \text{ in } \textit{TIS } | x_{p-1} \text{ at pos. } p-1 \text{ in } \textit{TIS })}{P(x_p \text{ at pos. } p \text{ in } \textit{ANY } | x_{p-1} \text{ at pos. } p-1 \text{ in } \textit{ANY })}$$

where x_p is the nucleotide incident at position p in the sequence corresponding to data point x, *TIS* is the set of training sequences centered around TIS, and *ANY* is the set of all training sequences.

Clearly, the key in the kernels of Zien *et al.* is the computation of the (complement of) Hamming distance between two sequences though they also consider local positional dependency. However, Hamming distance is not the best measure of dissimilarity in the comparison of (un-aligned) biomolecular sequences. First of all, it requires that the sequences have the same length. Second, in the processes of DNA replication and evolution, the errors like insertions and deletions (*i.e.* indels) of nucleotides are common. In particular, short tandem repeats (STR) are often hotspots for indels [Li and Graur, 1991]. [3] When indels are prevalent, the Hamming distance between two sequences is often an exaggerated over-estimation of the true dissimilarity.

On the other hand, the edit distance (also known as the *Levenshtein metric* [Levenshtein, 1966] and *evolutionary distance* [Sellers, 1974]) is a more general and accurate measure of sequence dissimilarities. The (basic, unweighted) edit distance between two sequences denotes the minimum number of edit operations that transform one sequence into the other. Typical edit operations include insertion, deletion and substitution, although other less-frequent operations such as transposition and

---

[3]This is important because the context of a TIS includes the 5' UTR, which usually contains STRs, for example CpG islands.

block moves can also be considered. For simplicity, we only consider the first three operations in this chapter. The edit distance between two sequences is a metric and tightly related to the optimal alignment between the two sequences.

In principle, one could perhaps find the TIS of an mRNA sequence by aligning the sequence with some mRNA sequences with known TISs and checking if an ATG codon is the TIS by comparing the edit distance with some threshold. However, such a naive approach (referred to as *template matching*) usually has a large error rate, especially for low-quality sequences (*e.g.* ESTs). Nevertheless, we observe that the above alignment approach uses a weak classifier (template matching) with a good dissimilarity measure (edit distance). On the other hand, Zien *et al.* use a sophisticated classifier (SVM) with a weak similarity measure (the complement of Hamming distance). This observation suggests that we might be able to recognize TISs more accurately by combining the merits of both approaches.

In order to incorporate the edit distance into SVMs, we define the edit kernel to measure the similarity between two sequences:

$$k(x,y) = e^{-\gamma \cdot edit(x,y)} \tag{2.8}$$

where $edit(x,y)$ is the edit distance between x and y, and γ is a positive real value to scale the kernel value for numerical stability. In fact, γ will play a very important role in making the kernel matrix positive definite, which we will discuss later. The edit kernel also has a natural probabilistic interpretation. Recall that the edit distance between two (biomolecular) sequences is the minimum number of the edit operations that transform one sequence into the other. Equivalently, we can think of such an edit process as a sequence of (independent) evolutionary events. However, these evolutionary events occur in nature with different probabilities. Let $P(b|a)$ denote the probability of nucleotide (or amino acid, as discussed later) a mutating into nucleotide (or amino acid) b, where a or b (but not both) could be a space (thus denoting an indel). Although identity substitutions of the form $a \rightarrow a$ are not used explicitly in string edit, to satisfy $edit(x,x) = 0$, let us assume that $P(a|a) = 1$. This assumption does not generally hold in molecular evolution, But it is usually not a serious problem in practice. Thus, the *cost* of an edit operation $a \rightarrow b$ [4] could be interpreted as the negative logarithm of the mutation (*i.e.* substitution or indel) probability $P(b|a)$ and the edit distance between two sequences could be interpreted as the summation [5] of the negative logarithms of the mutation probabilities:

$$edit(x, y) = -\sum_i \log P(y_i|x_i) \tag{2.9}$$

where $x_i \rightarrow y_i$ is the ith mutation in an optimal edit from x into y. We call (2.9) the *log probability* model. [6] Under this interpretation, the edit kernel defined in (2.8) is just the probability, raised to the power of γ, of sequence x mutating into y:

$$k(x, y) = \left(\prod_i P(y_i|x_i) \right)^{\gamma} \tag{2.10}$$

If two sequences are similar, the mutation probability $\prod_i P(x_i|y_i)$ will be large and thus the kernel value $k(x, y)$ will be large. Otherwise, the probability and the kernel value will be small. Therefore, such an edit kernel measures the similarity between two sequences in the sense of both evolution and probability. Note that, so far we have implicitly assumed that $P(a|b) = P(b|a)$, which may not always be true, especially for amino acids. When this assumption does not hold, we may interpret the edit distance as

$$edit(x, y) = -\frac{1}{2} \left(\sum_i \log P(x_i|y_i) + \sum_i \log P(y_i|x_i) \right) \tag{2.11}$$

which is the average of the negative log probability of transforming x into y and that of transforming y into x to keep the symmetry property.

---

[4]The cost is restricted to 1 or 0 in this basic string edit model. However, it will be relaxed to arbitrary nonnegative numbers in the next section.

[5]The additive cost scheme corresponds to the assumption that mutations at different sites occur independently. The assumption appears to be a reasonable approximation of the evolution of mRNAs, which are linear and unstructured. The cost scheme can be easily extended to accommodate more general probabilistic models such as affine gap costs.

[6]A similar model is the *log-odds ratio model* [Altschul, 1991, Durbin et al., 1998]. However, we think that the log probability interpretation is more direct and it perhaps deals with indels better.

To use the edit kernel in support vector machines, we would hope that it is positive definite. Unfortunately, it has been shown that the edit kernel is not positive definite [Cortes et al., 2003, Cortes et al., 2002]. However, we can still use the edit kernel in support vector machines according to the following theorem [Schölkopf, 1997].

Theorem 2.2 *Suppose the data $x_1, \ldots, x_\ell$ and the kernel $k(\cdot, \cdot)$ are such that the matrix*

$$K_{ij} = k(x_i, x_j) \tag{2.12}$$

is positive. Then it is possible to construct a map Φ into a feature space $\mathcal{F}$ such that

$$k(x_i, x_j) = \langle \Phi(x_i), \Phi(x_j) \rangle \tag{2.13}$$

Conversely, for a map Φ into some feature space $\mathcal{F}$, the kernel matrix $K_{ij} = \langle \Phi(x_i), \Phi(x_j) \rangle$ is positive.

This theorem implies that, even though the kernel $k(\cdot, \cdot)$ is not positive definite, we can still use $k(\cdot, \cdot)$ in support vector machines or other algorithms that require $k(\cdot, \cdot)$ to correspond to a dot product in some space if the kernel matrix K is positive for the given training data. In fact, Theorem 2.2 does not even require $x_1, \ldots, x_\ell$ to belong to a vector space. This is very useful for biological sequences analysis because it is hard to come up with a sensible vector representation of biomolecular sequences. Thus, we can use the edit kernel in SVMs if we could make the kernel matrix positive definite. This can be achieved by adjusting the parameter γ. It is well known that a symmetric diagonally dominant real matrix with nonnegative diagonal entries is positive definite. If we choose a large enough γ, we can make the kernel matrix diagonally dominant and thus positive definite. In fact, this requirement could be relaxed in practice. We found in our experiments that a reasonably large γ could already make the kernel matrix positive definite but not diagonally dominant.

Although the edit kernel in (2.8) has a good biological interpretation, it cannot be applied directly to predict TISs. First, the start codon ATG may be at different positions in different sequences. The above simple edit/alignment approach does not pay attention to the position of a putative start codon and the content of its neighbors (*e.g.* similarity to the consensus sequence GCCACCatgG). Second, the nucleotides in the 5' UTR upstream of a start codon and the ones in the downstream region (*i.e.* nucleotides encoding amino acids) usually follow different evolutionary processes because of the selective pressure exerted on protein-coding regions. Therefore, we should compute the edit distance for (putative) 5' UTRs and downstream regions separately. In other words, we should employ the edit kernel as follows:

$$k(x, y) = e^{-\gamma_1 \cdot edit(x',y') - \gamma_2 \cdot edit(x'',y'')} \tag{2.14}$$

where $\gamma_1, \gamma_2 > 0$; $x = x'x'', y = y'y''$ are two (mRNA) sequences (with postulated TISs); x', y' and x'', y'' are the putative 5' UTRs and downstream coding regions of x and y, respectively. We call the kernel in (2.14) *edit kernel I*. We can still easily make the kernel matrix of edit kernel I positive definite by adjusting γ_1 and γ_2 because the product of two positive definite matrices is still positive definite.

2.3 Redundancy of Codons and More General Edit Costs

We can improve edit kernel I by considering the following simple biological fact. The genetic code is highly *redundant* because there are 61 valid codons (including the ATG start codon) and only 20 amino acids exist [Freifelder, 1987]. Moreover, the redundancy is not random. In fact, certain features of the redundancy are quite regular. For example, pairs of codons of forms XYC and XYT always code for the same amino acid and pairs of forms XYG and XYA usually code for the same amino acid. Since the region of an mRNA downstream of the TIS codes for amino acids, it makes sense to consider such a downstream region as an amino acid sequence rather than a sequence of nucleotides when computing the edit distance $edit(x'', y'')$. So, we define a new kernel, *edit kernel II*, that has the same form as edit kernel I but with a different domain. The domain of edit kernel I is $\mathcal{X} \times \mathcal{X}$, where $\mathcal{X} = \{$all nucleotide sequences$\}$. However, the domain of edit kernel II is $\mathcal{X}' \times \mathcal{X}'$, $\mathcal{X}' = \{$all sequences of nucleotides and amino acids$\}$.

2.3.1   Edit Costs Based on Mutation Probabilities

In an attempt to further improve the above edit kernels, we generalize
the cost model of edit operations. Edit kernels I and II use the unit cost
model, *i.e.* the cost of editing a into b is 1 if $a \neq b$ or 0 otherwise. The
model's predominant virtue is its simplicity. In general, more sophisticated cost models should be used. For example, substitutions between
two purines (or pyrimidines), *i.e.* transitions, are more frequent than
those between a purine and a pyrimidine, *i.e.* transversions, and thus
should cost less according to the connection between edit distance and
mutation probabilities. Similarly, replacing an amino acid with a bio-
chemically similar one should cost less than replacements using amino
acids with totally different properties. We call the kernel defined on a
(weighted) edit distance using some general edit cost matrices for nucleotides and amino acids *edit kernel III*.

A general edit cost matrix can be defined for nucleotides based on
some fixed transversion/transition ratio. The most widely used (similarity) score matrices for amino acids are PAM [Dayhoff et al., 1978]
and BLOSUM [Henikoff and Henikoff, 1992] matrices. PAM matrices
are based on the Dayhoff model of evolutionary rates. Using an alternative approach, BLOSUM matrices were derived from about 2000 blocks
of aligned sequence segments characterizing more than 500 groups of related proteins. Although PAM and BLOSUM matrices are popular and
are good for sequence alignment, neither can be applied directly in the
edit kernel because they are generated following the log-odds ratio model
rather than the log probability model.

To obtain a cost matrix for edit kernel III, we propose the following
algorithm based on the log probability model:

1. *Raise the 1-PAM matrix to the power of p and denote it M;*

2. $M \leftarrow -\log M;$

3. *Calculate the average value m of the diagonal of M;*

21

4. $M \leftarrow M - m$;

5. Set the diagonal elements of M to zero;

6. If there are some negative elements in M, set them to zero;

where 1-PAM is a substitution matrix that describes the probability of a substitution in the unit time, *i.e.* the period during which 1% of a nucleotide (or amino acid) sequence is expected to change. The entries of 1-PAM are of the form $P(b|a, t = 1)$ denoting the probability of a mutating into b in the unit time. The algorithm works as follows. By raising 1-PAM to the power of p, *e.g.* $p = 120$ or 250, we obtain the matrix M with entries $P(b|a, t = p)$, which is the probability of a mutating into b in p time units. Then we take logarithm on M and change the sign to obtain a cost matrix that is suitable for sequences with an average divergence of p time units. In order to make it satisfy the condition $edit(x, x) = 0$, we subtract the average m of the diagonal from M. Such a shift makes the average of the diagonal of the cost matrix zero. Now we can set the diagonal to zero. [7] Usually, there may be a very small number of off-diagonal entries slightly less than zero after the shift. We also set them to zero to make the matrix nonnegative. We call the cost matrix calculated by this algorithm a *Substitution Cost Matrix* (SCM). Note that, an SCM may be asymmetric. When using such an asymmetric cost matrix, we need modify the edit distance definition by taking the average of $edit(x, y)$ and $edit(y, x)$ in the kernel as discussed in the previous section.

To obtain the cost matrices for nucleotides, we use the 1-PAM matrix from [Mount, 2001], which is based on Kimura's two-parameter (K2P) model of nucleotide substitutions [Kimura, 1980]. Specifically, the probability of a transition for each nucleotide is 0.006 and that of a transversion is 0.002. The cost matrix with $p = 250$ is listed in Table 2.1. Note that, the matrix is symmetric.

Since the 1-PAM matrix for amino acids is not symmetric [Dayhoff

---

[7]One may attempt to set the diagonal entries $-\log P(a|a, t = p)$ to zero directly without the shift. However, $P(a|a, t = p)$ is usually small when p is large and thus $-\log P(a|a, t = p)$ is not close to zero, So, setting the diagonal to zero directly might introduce significant unfair bias against non-identity substitutions and indels.

Table 2.1: The SCM250 cost matrix for nucleotides.

	A	C	G	T
A	0.0000	0.3009	0.0626	0.3009
C	0.3009	0.0000	0.3009	0.0626
G	0.0626	0.3009	0.0000	0.3009
T	0.3009	0.0626	0.3009	0.0000

et al., 1978], a resulting SCM for amino acids may not be symmetric. We may use such a matrix directly and modify the edit distance definition in the kernel to make it symmetric as mentioned above. Alternatively, we may make the resulting cost matrix symmetric by computing $M \leftarrow (M + M^T)/2$ either (i) immediately before step 6 or (ii) immediately before step 2. The former option seems more logical, but it might result in an edit distance similar to the above modified edit distance. Hence, we will only consider the latter option as an alternative, and call the cost matrix resulted from this approach an *Approximate Substitution Cost Matrix* (ASCM). An SCM cost matrix for amino acids with $p = 250$ is listed in Table 2.2.

Finally, we need define costs for indels. Since there are few rigorous treatments of indels, we define indel costs based on empirical experience. By some preliminary experiments, we have found that SVMs usually perform better if we set the indel cost for 5' UTRs small and that for the downstream region relatively large (the actual values will be given in the next section). This is consistent with the fact that the indels are expected to be more frequent in 5' UTRs than in downstream regions that code amino acids.

2.4 Experiments

To evaluate the performance of our algorithm for predicting TISs, we test all three edit kernels on Pedersen and Nielsen's original data set [Pedersen and Nielsen, 1997] and some small data sets derived from the data. The experimental results, as described below, show that our methods perform

Table 2.2: The SCM250 cost matrix for amino acids.

	Ala	Arg	Asn	Asp	Cys	Gln	Glu	Gly	His	Ile
Ala	0.0000	1.0288	0.6270	0.6013	1.1485	0.7606	0.5934	0.3688	0.9903	0.7640
Arg	1.8122	0.0000	1.4515	1.7407	2.2667	1.1464	1.6854	2.1189	1.0766	1.8287
Asn	1.3744	1.4431	0.0000	0.9558	2.2558	1.2619	1.1007	1.3277	1.0902	1.8317
Asp	1.1885	1.5994	0.8086	0.0000	2.4359	0.9167	0.5009	1.1236	1.1471	1.8011
Cys	2.1520	2.3747	2.4974	2.8536	0.0000	2.8506	2.8806	2.6636	2.4044	2.2314
Gln	1.6104	1.2289	1.3249	1.1284	2.7457	0.0000	0.9326	1.8024	0.8330	1.9517
Glu	1.1337	1.4899	0.9001	0.4475	2.4284	0.6629	0.0000	1.1732	1.0959	1.6573
Gly	0.3518	1.2276	0.5681	0.5141	1.4189	0.9065	0.6062	0.0000	1.1413	1.2020
His	1.9552	1.2860	1.2679	1.4856	2.4181	0.9739	1.5091	2.1935	0.0000	2.2691
Ile	1.6842	2.0071	1.9466	2.0938	1.9942	2.0105	2.0369	2.1497	2.1313	0.0000
Leu	1.1537	1.3987	1.3301	1.6096	2.0744	1.0983	1.4796	1.5850	1.1605	0.1232
Lys	1.0046	0.0000	0.5183	0.7172	1.9786	0.5806	0.7472	1.1485	0.7751	1.1524
Met	2.7194	2.5547	2.8212	3.0657	3.6201	2.6505	2.9833	3.1669	2.9342	1.9184
Phe	2.1427	2.3371	2.1122	2.6012	2.3679	2.4438	2.5744	2.3206	1.8045	1.2066
Pro	0.9213	1.2422	1.2908	1.3707	1.8339	1.1129	1.2793	1.2794	1.2333	1.6099
Ser	0.6327	0.9573	0.7302	0.8164	0.9170	0.9984	0.8803	0.6450	1.0993	1.2013
Thr	0.7754	1.2408	0.9567	1.0868	1.5666	1.2364	1.1424	1.0474	1.3877	1.0372
Trp	4.1465	2.3507	4.0147	4.5089	4.4423	4.1136	4.5235	4.4084	3.9310	4.2786
Tyr	2.4368	2.8054	2.1614	2.6514	1.6218	2.6273	2.6000	2.8303	1.7231	1.9500
Val	0.9590	1.5266	1.3838	1.4980	1.4263	1.4188	1.4346	1.3315	1.5029	0.1265

	Leu	Lys	Met	Phe	Pro	Ser	Thr	Trp	Tyr	Val
Ala	1.0608	0.9189	0.9299	1.4089	0.3837	0.4102	0.3822	2.0051	1.5063	0.6157
Arg	2.0535	0.6364	1.4917	2.2782	1.4798	1.5275	1.6300	0.9498	2.4702	1.9885
Asn	2.1256	1.1884	1.8476	2.1482	1.4958	1.2568	1.3134	2.3480	1.9161	1.8199
Asp	2.1670	1.2707	1.9026	2.4924	1.4231	1.1918	1.2914	2.8151	2.3072	1.7417
Cys	3.0225	2.8388	2.8337	2.6512	2.1350	1.6894	2.0654	3.4106	1.6243	2.0090
Gln	1.8786	1.3129	1.7412	2.4912	1.4180	1.6237	1.6749	2.6009	2.4778	1.9543
Glu	1.9149	1.2394	1.7237	2.3930	1.2870	1.2097	1.2900	2.8204	2.2546	1.6099
Gly	1.5027	1.0430	1.3131	1.7617	0.7339	0.3931	0.6308	2.2513	1.9164	0.9339
His	2.2386	1.6246	2.1971	1.9660	1.6870	1.8340	1.9248	2.2129	1.6833	2.2617
Ile	1.0006	1.9119	1.0465	1.3243	2.0545	1.8714	1.5128	2.7572	1.8350	0.6836
Leu	0.0000	1.2478	0.0000	0.2937	1.2701	1.3498	1.0543	1.1239	0.9640	0.2597
Lys	1.4263	0.0000	0.6409	1.8984	0.9669	0.7738	0.7315	1.5634	1.8331	1.2834
Met	1.5751	2.2624	0.0000	2.4703	2.9574	2.7928	2.5171	3.4405	3.1083	2.0059
Phe	1.0519	2.5536	1.3868	0.0000	2.5790	2.0818	2.0377	1.3594	0.0000	1.7456
Pro	1.6878	1.4042	1.6342	2.0554	0.0000	0.9767	1.1066	2.4848	2.3167	1.4815
Ser	1.5323	0.9043	1.2614	1.6077	0.6480	0.0000	0.5726	1.4826	1.6151	1.1126
Thr	1.4431	1.0372	1.2016	1.7432	0.9501	0.7444	0.0000	2.2463	1.7763	0.9951
Trp	4.4659	3.6745	4.2634	2.5395	4.1474	3.2582	3.9810	0.0000	2.5088	4.6172
Tyr	1.9193	2.9312	2.2830	0.1207	2.8663	2.3555	2.3499	1.7131	0.0000	2.2774
Val	0.5623	1.4650	0.5669	1.2317	1.2693	1.2151	0.9088	2.4074	1.5771	0.0000

significantly better than the Salzberg method, neural networks, and the SVM with Salzberg kernel. [8] In particular, the sensitivity and specificity of our methods are much higher than those of the previous methods. Just like accuracy, high sensitivity and specificity are both key desirable properties in a practical prediction application. In the following, we will describe Pedersen and Nielsen's data set, our experimental results, and some efficiency issues in the implementation of our algorithm.

2.4.1 Data

The test data consists of a collection of 3312 sequences from vertebrates, which were originally extracted from GenBank by Pedersen and Nielsen [Pedersen and Nielsen, 1997]. To mimic mRNAs, all sequences were "spliced" by removing possible introns and joining the remaining exon parts. Besides, only sequences containing at least 10 nucleotides upstream and at least 150 nucleotides downstream of their respective start codons were selected. A very thorough reduction of redundancy was performed to avoid over-estimating the prediction accuracy.

Because all the sequences contain TISs (*i.e.* they are all positive samples), we generate our test samples (both positive and negative) as follows [Pedersen and Nielsen, 1997,Zien et al., 2000]. For every potential start codon ATG in some sequence, a new sequence of length at most 210 nucleotides is extracted. Each of these sequences contains at most 30 nucleotides upstream and at most 180 nucleotides downstream (relative to the A in the putative start codon ATG). This leads to 13503 training and testing sequences, of which 3312 are positive and the rest are negative samples.

Note that in the experiments of Pedersen and Nielsen [Pedersen and Nielsen, 1997] and Zien *et al.* [Zien et al., 2000], they used a 200 nucleotide window approach to generate samples. Each window contains a potential

---

[8]Unfortunately, we were not able to obtain the program and data of Hatzigeorgiou and compare our method with her method in [Hatzigeorgiou, 2002]. The method of Davuluri *et al.* [Davuluri et al., 2001] requires the presence of the first introns and thus cannot be applied to Pedersen and Nielsen's data.

Table 2.3: Comparison of six-fold cross validation classification accuracies among different TIS prediction methods.

Method	Parameters [a]	Accuracy	Spec. [b]	Sens. [b]	Corr. [b]
Neural Network [c]		84.6%	64.5%	82.4%	62.7%
Salzberg method [d]		86.2%	73.7%	68.1%	61.9%
SVM [d] Salzberg kernel	$d_1 = 3$ $l = 1$	88.6%	76.0%	78.4%	69.6%
SVM edit kernel I	$C = 4$ $\gamma_1 = 0.00195$ $\gamma_2 = 0.00391$	93.2%	94.5%	89.3%	82.2%
SVM edit kernel II	$C = 16$ $\gamma_1 = 0.00017$ $\gamma_2 = 0.00049$	96.5%	96.0%	98.0%	91.0%

[a] The parameters were chosen by the five-fold cross validations on a small data set.

[b] Spec., Sens., and Corr. are specificity, sensitivity, and Mathews correlation coefficient, respectively.

[c] The results of neural network were obtained by Pedersen and Nielsen [Pedersen and Nielsen, 1997].

[d] The results of the Salzberg method and the SVM with Salzberg kernel were obtained by Zien *et al.* [Zien et al., 2000].

TIS at the center of the window. For an ATG codon near the (left or right) end of an mRNA sequence, each missing position in the test sequence (window) is filled with an 'N' (unknown). However, here we do not need fill in 'N's for missing positions because our kernels are based on string edit, which does not require the sequences have the same length.

2.4.2   Experimental Results

The experimental results are shown in Tables 2.3 and 2.4. All the results are based on the six-fold cross validation method, as done in [Pedersen and Nielsen, 1997] and [Zien et al., 2000]. Namely, the data is divided into

26

six parts of approximately equal sizes and each part is in turn reserved for testing the SVM learned on the other five parts. The average of the six prediction results are reported. The parameters C, γ_1, γ_2 are chosen optimally by cross validation experiments on a small data set of size 500. In the tables, accuracy is defined as $(TP + TN)/N$, specificity is $TN/(TN+FP)$, [9] sensitivity is $TP/(TP+FN)$, and Mathews correlation coefficient is defined as:

$$Corr. = \frac{TP \times TN - FP \times FN}{\sqrt{(TP + FP)(TP + FN)(TN + FP)(TN + FN)}}$$

where N, TP, TN, FP, and FN denote the numbers of the test samples, true positives, true negatives, false positives, and false negatives, respectively.

Using the SVM with edit kernel I, our algorithm performed reasonably well on the data. The accuracy (93.2%) is much better than that (88.6%) of the SVM with Salzberg kernel. In particular, the specificity (94.5%) and the sensitivity (89.3%) are significantly better than those (76.0% and 78.4%, respectively) of the SVM with Salzberg kernel. With edit kernel II, the accuracy is improved to 96.5%. The specificity is slightly improved to 96.0%, while the sensitivity is significantly improved to 98.0%.

In order to evaluate edit kernel III, we used the cost matrices SCM120, ASCM120, SCM250, and ASCM250, *i.e.* the SCM and ASCM with $p = 120$ and $p = 250$, respectively. Considering the popularity of PAM250, we also tested edit kernel III with PAM250. Of course, the PAM250 matrix cannot be used directly in the edit kernel because the edit distance based on PAM250 does not meet the requirements of metric. Thus, we modify PAM250 as follows. We first convert PAM250 into a cost matrix. Then we shift PAM250 up by subtracting the average of the diagonal from every element in the matrix. Finally, we set the diagonal elements and the two off-diagonal negative elements to zero. The comparative results of edit kernel III with these different cost matrices are shown in Table 2.4, which are in general much better than those in Table 2.3. Again, the parameters and gap penalties were chosen by experiments on a small data set. The SVMs with matrices SCM120, SCM250, ASCM120, and ASCM250 all

---

[9]Zien *et al.* used a different definition, $TP/(TP + FP)$, which is usually called *precision* [Han and Kamber, 2000].

Table 2.4: Comparison of six-fold cross validation classification accuracies of SVMs with edit kernel III using different cost matrices.

Cost Matrix	Parameters	Indel [a] Penalty	Accuracy	Spec.	Sens.	Corr.
SCM120	$C = 32$ $\gamma_1 = 0.00195$ $\gamma_2 = 0.00195$	1.00 9.00	99.64%	99.73%	99.37%	99.02%
SCM250	$C = 16$ $\gamma_1 = 0.00195$ $\gamma_2 = 0.00781$	0.35 8.00	99.84%	99.88%	99.73%	99.58%
ASCM120	$C = 8$ $\gamma_1 = 0.00195$ $\gamma_2 = 0.00781$	1.00 9.00	99.67%	99.69%	99.61%	99.10%
ASCM250	$C = 8$ $\gamma_1 = 0.00195$ $\gamma_2 = 0.00781$	0.35 7.00	99.90%	99.92%	99.82%	99.72%
PAM250	$C = 8$ $\gamma_1 = 0.00195$ $\gamma_2 = 0.00195$	0.35 13.9	97.75%	98.17%	96.44%	93.97%

[a] In the indel penalty column, the first number is the penalty for nucleotides and the second is the penalty for amino acids.

performed equally well, with only some minor differences. The accuracy is amazingly improved to 99.9% with the cost matrix ASCM250. On the other hand, the accuracy of the SVM with PAM250 is 97.7%, which is only slightly better than that of the SVM with edit kernel II employing the unit cost. This could perhaps be an evidence suggesting that the log probability model is more suitable for the edit distance in edit kernel III than the log-odds ratio model. On the other hand, the results in Table 2.4 also indicate that the SVM with edit kernel III is not very sensitive to the actual cost matrix used as long as it is based on the log probability model (although it might require different parameters for different matrices).

In fact, the SVM with edit kernel III is better than it appears in Table 2.4, which is based on cross validations on a large data set with 13503 samples. On small data sets, the SVM with edit kernel III can

Table 2.5: Comparison of the accuracies [a] among the edit kernels on small data sets of size 500, 1000, and 2000.

Kernel	500	1000	2000
Edit kernel I	86.0%	86.4%	89.5%
Edit kernel II	92.0%	92.9%	95.1%
Edit kernel III [b]	98.6%	99.3%	99.4%

[a] The accuracy is estimated by six-fold cross validations.
[b] Edit kernel III employs the SCM250 cost matrix.

Table 2.6: Comparison of the average numbers of support vectors among the edit kernels.

Kernel	Average number of SVs
Edit kernel I	2312
Edit kernel II	2316
Edit kernel III, SCM120	319
Edit kernel III, SCM250	230
Edit kernel III, ASCM120	507
Edit kernel III, ASCM250	293
Edit kernel III, PAM250	821

still achieve a very high accuracy, as shown in Table 2.5. In contrast, the accuracies of the SVMs with edit kernels I and II drop notably on the small data sets. For example, on a set of 500 samples, the SVM with edit kernel III based on SCM250 can achieve an accuracy of 98.6%, while the SVMs with edit kernels I and II could only achieve 86.0% and 92.0%, respectively.

Besides accuracy, it is also important to compare the SVMs with different edit kernels in terms of the number of support vectors. It is known that the set of support vectors provides a "compressed" version of the training data containing all the information necessary to solve a given classification task [Schölkopf, 1997]. The average number of support vectors of the SVMs with edit kernels I, II and III on the large set of 13503 samples are listed in Table 2.6. These numbers are again based on six-fold cross validations. The SVMs with edit kernel I and II have about 2300 support vectors, which is pretty small compared with the number

Table 2.7: Comparison of the average numbers of iterations in the training of SVMs with different edit kernels.

Kernel	Average number of iterations
Edit kernel I	105065
Edit kernel II	20208
Edit kernel III [a]	4466

[a] Edit kernel III employs the SCM250 cost matrix.

of training samples. Moreover, the SVM with edit kernel III based on SCM250 uses only 230 support vectors. In contrast, the SVM based on SCM120 has 319 support vectors. This indicates that SCM250 (*i.e.* the substitution cost matrices over 250 evolutionary time units) is perhaps more suitable than SCM120 for Pedersen and Nielsen's data set of vertebrate mRNAs. Besides, when ASCM250 and ASCM120 are used, the SVMs require 293 and 507 support vectors, respectively. A reason that slightly more support vectors are required here is probably because some information is lost when we take the average on the probabilities in an ASCM. The fact that the SVM with PAM250 requires 821 support vectors suggests again that the log-odds ratio model may not be as suitable as the log probability model for the edit kernel.

A reason that the edit kernel III uses a small number of support vectors may be that evolutionary information is already incorporated into the kernel so that the SVM is able to do prediction with fewer support vectors. It is also known that the expectation of the number of learned support vectors from a training set of size ℓ, divided by $\ell - 1$, is an upper bound on the expected probability of test error [Vapnik and Chervonenkis, 1974]. Thus, the very small number of support vectors required by edit kernel III with SCM250 is an assurance of the SVM's good performance.

2.4.3 Computational Issues

The time of training an SVM depends on both the number of iterations and the time complexity of the kernel function. If the *sequential mini-*

mal optimization (SMO) is employed to train the SVM, the number of iterations ranges somewhere between linear and quadratic in the training set size [Platt, 1999], depending on the actual data and kernel function. Table 2.7 lists the average numbers of the iterations used in training the SVMs with different edit kernels in the above six-fold cross validation experiment on the large data set. For all three edit kernels, the numbers of iterations are (roughly) proportional to the size of training data, although the SVM with edit kernel I seems to require significantly more iterations.

In the prediction/testing phase, the speed of an SVM depends on both the number of support vectors and the time complexity of the kernel function. According to the indicator function (2.4), the fewer support vectors, the faster the SVM predicts. Since the SVM with edit kernel III requires the fewest support vectors as shown in Table 2.6, they are the fastest in the prediction phase as well.

In both training and testing phases, the time complexity of the kernel function plays an important role in the speed of an SVM. Many kernel functions, *e.g.* the polynomial kernels, can be computed in $O(n)$ time, where n is the length of input vectors (or sequences). However, our edit kernels have time complexity $O(n^2)$ based on dynamic programming. To improve the time complexity, one may attempt to use some fast algorithm to compute the edit distance, such as Ukkonen's algorithm [Ukkonen, 1985]. Ukkonen's algorithm runs in $O(s * n)$ time for instances of length n and edit distance s. Ukkonen's algorithm, however, works only for the unit cost model. Although Ukkonen's algorithm has been improved by Wu *et al.* [Wu et al., 1990, Wu and Manber, 1992] and Berghel and Roach [Berghel and Roach, 1996], the improved algorithms still depends on the unit cost model. Therefore, Ukkonen's algorithm and its improvements are not suitable for edit kernel III. Finding a fast (approximate) algorithm for edit kernel III is an interesting future research topic. For example, we may limit the number of indels, say k, so that we need only compute the k-diagonal elements in the table of dynamic programming. Such a band-based algorithm is particularly suitable for computing edit distances on coding regions since indels are usually not frequent in these regions.

2.4.4 Discussion

The above experimental results show that the SVMs with the edit kernels have a very high accuracy in predicting TISs that is unmatched by any previous technique. Why do these algorithms work so well? Below, we try to give some insights from several perspectives.

First of all, our algorithms are based on SVMs. SVMs have a solid foundation in statistical learning theory and have proven to be more general and powerful than many other learning techniques in various applications. More precisely, the capacity control capability makes SVMs free of the overfitting problem [Vapnik, 1995, Vapnik, 1998]. SVMs can also been interpreted in the framework of regularization theory [Tikhonov, 1963], which is a general approach to handle ill-posed problems. The small number of support vectors used in an SVM also has natural interpretations in the context of Kolmogorov complexity and minimum description length (MDL) principle [Vapnik, 1995, Li and Vitányi, 1997].

Although SVMs are general and powerful, they are not a silver bullet. The performance of an SVM largely depends on the choice of its kernel. In general, the more prior knowledge is incorporated into the kernel function, the better the performance of the SVM. This is why the SVM with Salzberg kernel that incorporates local dependency in the sequences performs better than the SVM with a plain polynomial kernel based on Hamming distance [Zien et al., 2000]. Our edit kernel I is based on the edit distance, which is more general than the Hamming distance and measures the dissimilarity between sequences more accurately. With such a simple improvement on the kernel, our SVM achieved a notably higher accuracy than the SVM with Salzberg kernel. Therefore, it is natural to expect that the SVMs with edit kernels II and III, which take into account more prior knowledge including codon redundancy and evolutionary costs between nucleotides and amino acids, would perform even better than the SVM with edit kernel I.

Table 2.8: Six-fold evaluation of TISHunter with the SCM250 cost matrix on Pedersen and Nielsen's data set of 3312 sequences.

Method	Sample size	Error
ESTScan, closest ATG [a]	2350	729
Salzberg method [a]	3312	1095
SVM, Salzberg kernel [a]	3312	530
TISHunter	3312	13

[a] The results of ESTScan, the Salzberg method and the SVM with Salzberg kernel were obtained by Zien *et al.* [Zien et al., 2000].

2.5 An Online Program: TISHunter

Based on the SVM with edit kernel III, we have developed an online program *TISHunter* for the prediction of TISs in mRNA sequences, which is publicly accessible at http://bioinfo.ucr.edu/~hli. TISHunter uses SCM250 because it has resulted in the high accuracy, the smallest number of support vectors, and the fastest training time in the tests done so far. TISHunter works on a "per sequence" basis, *i.e.* it linearly scans each input mRNA sequence and predicts a score for every potential start codon ATG. The ATG codon with the largest predicted value is output as the putative TIS in the sequence if the score is positive. Note that TISHunter does not assume that the input sequence must have a TIS.

To evaluate TISHunter, we again divide Pedersen and Nielsen's original set of 3312 sequences into six parts, each of which has 552 sequences. For each part, we train TISHunter on the other five parts and test it on the missing part. The test results and comparisons with those of the previous methods are summarized in Table 2.8. Note that such a comparison is not fair for ESTScan since it was not trained on these sequences. Also, Zien *et al.* took advantage of the fact that input every input mRNA sequence has a TIS, and simply chose the ATG codon with the largest predicted value as TIS even if the value is negative.

It is amazing that TISHunter made only 13 incorrect predictions totally, while the other programs made at least 40 times more errors. It turns out that these 13 errors are all false negatives (*i.e.* no TISs were predicted in the involved sequences) because none of the candidate start codons in the sequences received a positive prediction score. If we take advantage of the fact that these mRNA sequences all have TISs, we could modify TISHunter to simply output the ATG codon with the largest prediction score as the putative TIS in each sequence. Interestingly, the modified program would make no errors on this data set. The reason that the modified program could perform perfectly here is that, when working on a "per ATG codon" basis, our algorithm usually gives a true start codon a larger score than those of the true negatives in the same mRNA sequence. Thus, even if a true start codon is incorrectly predicted as a negative, its prediction score could still be larger than those of the true negatives in the same sequence and will thus be output as a putative TIS by the modified program.

2.6   Prediction on the Human mRNAs

We downloaded all human mRNAs with the status code REVIEWED from NCBI Reference Sequence (RefSeq) database [Pruitt and Maglott, 2001]. These sequences have been reviewed by NCBI staffs or their collaborators and we may assume that they are of high quality. The dataset contains 8824 sequences. After deleting the sequences whose upstream region to the left of TIS is less than 10 nucleotides or downstream region is less than 150 nucleotides, we kept 8225 sequences as the experimental dataset. These 8225 sequences have an average length of 2844 and contain 417880 potential ATG start codons. Of the 8225 sequences, there are only 4400 (53.5%) sequences that use the first ATG as TIS. It is very different from Kozak's claim that downstream ATGs are used as start codons in less than 10% of her investigated eukaryotic mRNAs [Kozak, 1989].

Because the signal-to-noise ratio $(8225/409655 = 2\%)$ is very low, we use a different experimental setting from what we used on Pedersen and Nielsen's data. For each ATG before (and including) the true TIS, we generate a data point that contains at most 30 nucleotides upstream and 270 nucleotides downstream. This length is determined based on our experiments on small size data. Note that the downstream region is longer than that used on Pedersen and Nielsen's eukaryotic mRNAs. With this setting, we generate 20877 data points, of which 8825 are positive. Based on a three-fold cross validation, the support vector machine achieves 96.7% accuracy with edit kernel III and SCM250 cost matrix. The specificity, sensitivity, and correlation are 95.7%, 98.0%, and 93.1%, respectively. The average number of support vectors are 4961, which are 23.8% of the training data. In the experiment, the parameter C, γ_1, and γ_2 are set to 32, 0.0625, and 0.015625, respectively.

We also run TISHunter on the human mRNAs. TISHunter works on a per sequence basis and linearly scans the input mRNA sequence. The first ATG that gets a positive score is reported as the putative TIS. We use this setting because the training data contains only the ATGs before (and including) TISs. TISHunter can predict 92.9% of the 8225 true TISs correctly. Among the erroneously predicted ATGs, we find the following results. In the 584 incorrect predictions, 486 (83.2%) are in the upstream of the true TISs, 98 (16.8%) are in the downstream, 365 (62.5%) are in the reading frames, and 219 (37.5%) are out of the reading frames.

Interestingly, we find that, out of the 486 false predictions in the upstream of TISs, 173 (35.6%) contain a stop codon between the wrongly predicted position and the true start codon. We think that many of these false TISs may contribute to the procedure reinitiation. As we mentioned before, reinitiation happens when an 80S ribosome translates the first small open reading frame (upORF) and reaches a stop codon. In the study of human immunodeficiency virus type 1 mRNAs, Luukkonen *et al.* found that downstream translation initiation is inhibited in 50% of the cases by an upORF of 84 nucleotides and should be entirely abrogated by an ORF longer than 165 nucleotides (predicted by extrapolation) [Luukkonen et al., 1995]. In our 173 cases, 51 (29.5%) and 74 (42.8%) of the false TISs have stop codons in the 84 and 165 nucleotide downstream regions, respectively. Besides, reinitiation in eukaryotes is most efficient when the upORF terminates at some distance before the start of the next cistron [Kozak, 1987c]. The reason is that the 40S ribosomal subunit requires time (distance) to re-acquire Met-tRNA$_i$·eIF-2, without which the downstream ATG codon cannot be recognized [Hinnebusch, 1997]. According to the study of Luukkonen *et al.*, an intercistronic distance shorter than 37 nucleotides appears to negatively affect initiation frequency at downstream ATGs [Luukkonen et al., 1995]. In the 74 false TISs that have stop codons in the 165 nucleotides downstream region, 53 (71.6%) have the intercistronic distance larger than 37 nucleotides. These 53 false TISs may be related with the procedure reinitiation.

2.7   Conclusion

Automating the process of sequence annotations is an important part of the post-sequencing genomics research. In this chapter, we show that a powerful machine learning technique, SVMs, can effectively find TISs if we carefully incorporate the biological knowledge into the kernel function. The proposed method can be extended to the recognition of other biological signals, *e.g.* binding sites of regulatory proteins, because one can incorporate motif information into edit kernels. We hope that further analysis on the learned support vectors will elucidate the real mechanisms involved in a translation initiation process (*e.g.* how much does the ribosome care about a downstream region).

Chapter 3

Accurate Cancer Diagnosis with Gene Expression Profiling

3.1 Introduction

Accurate diagnosis of human cancer is essential in cancer treatment. Recently, the advances in microarray technology enable us to simultaneously observe the expression levels of many thousands of genes. In principle, tumor gene expression profiles can serve as molecular fingerprints for cancer classification. Researchers believe that gene expression profiling could be a precise, objective, and systematic method for cancer classification [Golub et al., 1999, Khan et al., 2001, Ramaswamy et al., 2001]. Many classifiers have been applied to cancer classification, such as nearest neighbor, artificial neural networks, support vector machines, boosting, weighted voting, etc. [Ben-Dor et al., 2000, Dettling, 2004, Dettling and Bühlmann, 2003, Dudoit et al., 2002, Furey et al., 2000, Golub et al., 1999, Khan et al., 2001, Mukherjee et al., 1998, Ramaswamy et al., 2001, Slonim et al., 2000, Yeang et al., 2001, Zhang et al., 2003].

Although gene expression profiling provides a great opportunity for accurate and objective cancer diagnosis, the classification task in this context is very challenging because of the very high dimensionality of data since gene expression data usually involve thousands of genes. The high dimensionality is a major practical limitation facing many pattern recognition technologies, especially when the number of samples is small. In practice, it has been observed that a large number of features may degrade the performance of classifiers if the number of training samples is small relative to the number of features [Jain and Chandrasekaran, 1982, Raudys and Jain, 1991, Raudys and Pikelis, 1980]. This fact, which is referred to as the "peaking phenomenon", is caused by the "curse of dimensionality" [Bellman, 1961]. It is generally accepted that one needs at least $5 - 10$ times as many training samples per class as the number of features to obtain well-trained (robust) classifiers [Foley, 1972, Jain and Chandrasekaran, 1982, Raudys and Jain, 1991]. In the case of cancer classification, the number of tumor samples is usually only several dozen due to limitations on sample availability, identification, acquisition, time, and cost. On the other hand, the dimensionality (number of genes) is many thousands. Consequently, dimensionality reduction is essential to cancer classification.

In the past several decades, many dimension reduction techniques have been proposed. Roughly, these methods follow two approaches, *feature selection* and *feature extraction*. Feature selection methods choose a "best" subset of features from a large initial set, taking into account both the cost of selection and the effectiveness of each feature in the classification process. The advantage of feature selection is that the selected features retain their original biological or physical interpretation. So, the retained features help us understand patterns more precisely, and even find the biological/physical process that generates the patterns. On the other hand, feature extraction techniques transform the original feature space into a reduced feature space that has fewer dimensions with little reduction on the effectiveness of classifier. Feature extraction generally provides a better discriminative ability than feature selection. However, the new features generated by (nonlinear) feature extraction may not have a clear biological/physical meaning.

Currently, almost all gene-expression-based cancer classification systems employ some feature selection method for dimension reduction. Ideally, one should select the expression levels of tumor-specific genes for classification. However, only few tumor-specific molecular markers are currently known by molecular oncology. Instead, we have to select marker genes computationally. Of the applied feature selection methods, the single-gene-rank approach is the most common. This approach ignores the interdependence among genes and ranks the relative class separation of each feature independently. The top ranked genes are selected for training the classifier. For example, Golub *et al.* used the signal-to-noise (S2N) metric ratio $P(g) = \frac{|\mu_1 - \mu_2|}{\sigma_1 + \sigma_2}$ to rank genes, where μ_1 and μ_2 are the mean expression levels of gene g in classes 1 and 2, respectively, and σ_1 and σ_2 are the standard deviations of expression levels in classes 1 and 2, respectively [Golub et al., 1999]. Of course, such univariate feature selection methods are not optimal because a subset of top k ranked genes is not guaranteed to be the best among all subsets of k genes. In particular, many genes are coexpressed due to the complicated genetic networks. Thus, the expression of genes is not independent and it is very hard for univariate feature selection methods to capture the joint discriminant capability of genes.

To improve single-gene-ranking, Bø and Jonassen suggested gene-pair-ranking that simultaneously analyzes pairs of genes to decide the subset of marker genes [Bø and Jonassen, 2002]. Clearly, it is not sufficient to investigate the joint discriminant capability of only a pair of genes. More generally, subsets of $K > 2$ features should be considered. However, it is not practical to jointly select a subset of genes in a brute-force way. For example, the number of ways to select 50 elements from 2000 elements is approximately 10^{100}. Instead, some heuristic has to be applied. In [Li et al., 2001a, Li et al., 2001b], for instance, Li *et al.* proposed the GA/KNN method for gene selection. First, GA/KNN finds many (random) subsets of $K > 2$ genes of expected classification power using a Genetic Algorithm (GA). The "fitness" of each subset of genes is determined by its ability to classify the training set samples according to the k-nearest neighbor (kNN) method. When many such subsets of genes are obtained, the frequencies with which genes are selected are analyzed. The most frequently selected genes are presumed to be the most relevant to sample distinction and are finally used for prediction. Although GA/KNN avoids brute-force search, it is still much slower than univariate feature selection and our proposed method (to be discussed later).

The user also has to determine many parameters of the algorithm, such as chromosome length, the number of chromosomes, termination metric, etc.

Due to its popularity and success in many application areas, some researchers have used support vector machine (SVM) to select features. For example, the one-dimensional SVM method ranks genes by the accuracies of single-gene SVM classifiers [Su et al., 2003]. This is actually an univariate feature selection method and does not exhibit superiority to other univariate methods [Li et al., 2004]. Note that, any other classifiers may be employed instead of SVM in this approach. Another feature selection method based on SVM is recursive feature elimination (RFE) [Ramaswamy et al., 2001]. RFE recursively removes features based on the absolute magnitude of the hyperplane elements of trained SVM. In RFE, the SVM is trained with all genes at first. The expression values of genes whose absolute value of corresponding hyperplane element is in the bottom 10% are removed. Then, the SVM is retrained with the selected genes. This procedure is repeated iteratively to study prediction accuracy as a function of the gene number. It was reported that RFE achieves the similar results as the signal-to-noise ratio method [Golub et al., 1999] and radius-margin-ratio method [Mukherjee et al., 1998, Weston et al., 2000] on the GCM dataset [Ramaswamy et al., 2001].

It was observed that no matter what feature selection method is employed, at least 50 (and frequently more) features would need be chosen and used for classification in general [Somorjai et al., 2003]. This number is quite far from the recommended 5 – 10 times ratio of samples to features per class for the training of a robust classifier, and thus the estimated error rate could be greatly biased [Foley, 1972, Jain and Chandrasekaran, 1982, Raudys and Jain, 1991]. Hence, it is not clear how well the previously proposed methods perform if large datasets are available.

Besides feature selection methods, some feature extraction methods, e.g. principal component analysis (PCA), had been applied for dimensionality reduction in cancer classification [Khan et al., 2001]. PCA is a linear mapping that minimizes the mean squared error criterion [Jolliffe, 1986]. PCA computes the d largest eigenvectors of the $D \times D$ covariance matrix of n D-dimensional samples. The d largest eigenvectors that constitute the mapping matrix are also called as the principal components. These few principal components describe most of the variance of the data. By expanding the data on these orthogonal principal components, we have the minimal reconstruction error. However, the decorrelation and high measures of statistical significance provided by the first few principal components are no guarantees of revealing the class structure that we need for proper classification. As done in many face recognition systems [Turk and Pentland, 1991], d is usually set to $n-1$ for no information loss when the sample size is much smaller than the dimensionality, where n is the number of samples. However, such a setting cannot meet the recommended 5 – 10 times ratio of samples to features per class. Besides, the fact that the category information associated with samples is neglected also implies that PCA may be significantly sub-optimal.

In this chapter, we will develop a robust and accurate system for dimension reduction in cancer classification based on linear discriminant analysis (LDA). After dimension reduction by LDA, a template matching procedure is employed for classification. LDA can map the data into the discriminant space with a very low dimensionality of $c-1$, where c is the number of classes. For instance, the mapped space is one dimensional for binary classification. So, the mapped data meet the recommended 5 – 10 times ratio of samples to features per class and thus even a small number of samples are sufficient to train a good classifier. Therefore, our method is more robust than others and the estimated error rates are more accurate and trustable. Although the method sounds straightforward, there is a big challenge. Namely, the conventional LDA cannot be applied when the within-scatter matrix $\mathbf{S}_w$ is singular due to the *small sample size problem* [Fukunaga, 1990]. The small sample size problem arises when the number of samples is smaller than the dimensionality of samples, which always happens in cancer classification. To overcome the small sample size problem, we developed the generalized linear discriminant analysis (GLDA), which is a direct improvement of LDA to overcome the small sample size problem by carefully investigating the properties of scatter matrices. GLDA is mathematically well-founded and coincides with the conventional LDA when $\mathbf{S}_w$ is nonsingular. Different from the conventional LDA, GLDA does not assume the nonsingularity of $\mathbf{S}_w$, and thus naturally solves the small sample size problem. To deal with the high dimensionality of scatter matrices, we also develop fast algorithms for GLDA based on singular value decomposition (SVD). Our experi-

41

mental results show that the method performs well. Especially on some difficult instances that have very small samples to genes per class ratios, our method achieves the much higher accuracies than the widely used classification methods such as support vector machines, random forests, etc.

The rest of the chapter is organized as follows. Section 3.2 introduces LDA and the small sample size problem. In Section 3.3, GLDA will be developed. Section 3.4 describes some experimental results on several public cancer gene expression datasets in comparison with many other methods. Section 3.5 concludes the chapter with some directions of further research.

3.2   Linear Discriminant Analysis

Linear discriminant analysis (LDA, also called Fisher's Linear Discriminant) is a supervised feature extraction (and classification) method [Fisher, 1936, Rao, 1948]. In many applications, LDA has proven to be very powerful. Besides, LDA can map the data into the discriminant space of dimensionality $c - 1$ so that we can meet the recommended $5 - 10$ times ratio of samples to features per class. Thus, we can classify tumors more robustly. Recall that LDA is given by a linear transformation matrix $\mathbf{W} \in \mathcal{R}^{D \times d}$ maximizing the so-called Fisher criterion (a kind of *Rayleigh* coefficient) [Fisher, 1936, Rao, 1948]

$$J(\mathbf{W}) = \mathrm{tr} \left(\frac{\mathbf{W}^T \mathbf{S}_b \mathbf{W}}{\mathbf{W}^T \mathbf{S}_w \mathbf{W}} \right) \tag{3.1}$$

where $\mathbf{S}_b = \sum_{i=1}^{c} p_i (\mathbf{m}_i - \mathbf{m})(\mathbf{m}_i - \mathbf{m})^T$ is the between-class scatter matrix and $\mathbf{S}_w = \sum_{i=1}^{c} p_i E[(\mathbf{x} - \mathbf{m}_i)(\mathbf{x} - \mathbf{m}_i)^T | \mathcal{C}_i] = \sum_{i=1}^{c} p_i \mathbf{\Sigma}_i$ is the within-class scatter matrix; c is the number of classes; $\mathbf{m}_i$ and p_i are the mean vector and *a priori* probability of class i, respectively; $\mathbf{m} = \sum_{i=1}^{c} p_i \mathbf{m}_i$ is the overall mean vector; $\mathbf{\Sigma}_i$ is the covariance matrix of class i; D and d are the dimensionalities of the data before and after the transformation, respectively; and tr denotes the trace of a square matrix, i.e. the sum of the diagonal elements. Besides, the total/mixture scatter matrix, i.e. the covariance matrix of all samples regardless of their class assignments, is defined as $\mathbf{S}_t = E[(\mathbf{x}-\mathbf{m})(\mathbf{x}-\mathbf{m})^T] = \mathbf{S}_w + \mathbf{S}_b$ [Fukunaga, 1990]. In LDA, the transformation matrix $\mathbf{W}$ is constituted by the largest eigenvectors of $\mathbf{S}_w^{-1}\mathbf{S}_b$ to maximize (3.1) by assuming the nonsingularity of $\mathbf{S}_w$. Since $\mathbf{S}_w^{-1}\mathbf{S}_b$ has at most $d = \min(c-1, D)$ non-zero eigenvalues, $\mathbf{W}$ is usually constituted by the corresponding eigenvectors $\mathbf{w}_1, \ldots, \mathbf{w}_d$. For cancer classification, d is at most $c-1$ since the number of classes is always less than the dimensionality. For a new sample $\mathbf{x}$, the predicted class is

$$\mathcal{C}(\mathbf{x}) = \arg\min_k \sum_{i=1}^{d} (\mathbf{w}_i^T(\mathbf{x} - \mathbf{m}_k))^2 \qquad (3.2)$$

i.e. the class whose mean vector is closest to $\mathbf{x}$ in the discriminant space.

As a linear feature extraction method, LDA could also be used to identify important marker genes for further investigation. After training, the elements of each column vector in the linear mapping matrix of LDA can be thought of as the weights of genes, which determine the importance of genes in classification. By ranking the absolute values of the elements of column vectors in the mapping matrix, we can select marker genes that have the highest ranks. These selected genes may be useful to find tumor-specific molecular markers. Moreover, our method considers the correlation among all genes and uses all genes to train the classifier, which is different from univariate feature selection. Thus, the selected marker genes may be more biologically meaningful.

3.2.1 The Small Sample Size Problem

Although LDA performs well in many applications, we cannot apply LDA directly when is $\mathbf{S}_w$ singular. Recall that, in order to find a $\mathbf{W}$ maximizing (3.1), we take the derivative of (3.1) with respect to $\mathbf{W}$ [Fukunaga, 1990],

$$\begin{aligned}
&\frac{\partial}{\partial \mathbf{W}} \mathrm{tr}((\mathbf{W}^T\mathbf{S}_w\mathbf{W})^{-1}(\mathbf{W}^T\mathbf{S}_b\mathbf{W})) \\
&= -2\mathbf{S}_w\mathbf{W}(\mathbf{W}^T\mathbf{S}_w\mathbf{W})^{-1}(\mathbf{W}^T\mathbf{S}_b\mathbf{W})(\mathbf{W}^T\mathbf{S}_w\mathbf{W})^{-1} \\
&\quad + 2\mathbf{S}_b\mathbf{W}(\mathbf{W}^T\mathbf{S}_w\mathbf{W})^{-1}
\end{aligned} \qquad (3.3)$$

Equating (3.3) to zero, an optimal $\mathbf{W}$ must satisfy

$$\mathbf{S}_b\mathbf{W} = \mathbf{S}_w\mathbf{W}(\mathbf{W}^T\mathbf{S}_w\mathbf{W})^{-1}(\mathbf{W}^T\mathbf{S}_b\mathbf{W}) \qquad (3.4)$$

If $\mathbf{S}_w$ is nonsingular, we can multiply its inverse to both sides of Equation (3.4) and obtain the conventional LDA through simultaneous diagonalization of $\mathbf{S}_w$ and $\mathbf{S}_b$ [Fukunaga, 1990]. When $\mathbf{S}_w$ is singular, however, this procedure cannot be applied.

Note that the rank of $\mathbf{S}_w$ is less than $n - c$ [Fukunaga, 1990], where n is the number of samples and c is the number of classes. So, $\mathbf{S}_w$ would be singular if the number of samples minus the number of classes is smaller than the dimensionality of the samples, which is referred as to the small sample size problem [Fukunaga, 1990]. This situation always happens in cancer classification since the data contains several thousand genes but only a few dozen.

3.2.2   Previous Work

In recent years, many researchers have noticed this problem and tried to overcome the computational difficulty with LDA. A simple and direct attempt is to replace $\mathbf{S}_w^{-1}$ with the pseudo-inverse matrix $\mathbf{S}_w^{+}$ [Tian et al., 1986]. However, it does not guarantee that Fisher's criterion is still optimized by the largest eigenvectors of $\mathbf{S}_w^{+}\mathbf{S}_b$. In fact, it can be shown that this approach produces very poor results (see the section of experiments). Another approach is to first reduce the dimensionality with some other feature selection/extraction method and then apply LDA on the dimensionality-reduced data. For instance, Belhumeur *et al.* proposed the Fisherface (also called PCA+LDA) method which first employs the principal components analysis (PCA) to reduce the dimensionality of the feature space to $n - c$, and then applies the standard LDA to reduce the dimensionality to $c - 1$, where n is the number of samples and c is the number of classes [Belhumeur et al., 1997]. Note that this method is suboptimal because PCA has to keep $n - 1$ principal components in order not to lose information. However, the first step of PCA+LDA keeps only $n - c$ principal components. Such a setting will lose too much information if the number of classes is large. Besides, it is not necessary to perform

44

the whole procedure of LDA after PCA is performed. As discussed later, we only need to calculate the largest eigenvectors of $\mathbf{S}_b$ after reducing the dimensionality of the feature space to $n-1$ with PCA.

To handle the singularity problem, it is also popular to add a singular value perturbation to $\mathbf{S}_w$ to make it nonsingular [Hong and Yang, 1991]. A similar but more systematic method is regularized discriminant analysis (RDA) [Friedman, 1989]. In RDA, one tries to obtain more reliable estimates of the eigenvalues by correcting the eigenvalue distortion in the sample covariance matrix with a ridge-type regularization. Besides, RDA is also a compromise between LDA and QDA (quadratic discriminant analysis), which allows one to shrink the separate covariances of QDA towards a common covariance as in LDA. Penalized discriminant analysis (PDA) is another regularized version of LDA [Hastie and Tibshirani, 1995, Hastie et al., 1994]. The goals of PDA are not only to overcome the small sample size problem but also to smooth the coefficients of discriminant vectors for better interpretation. In PDA, $\mathbf{S}_w$ is replaced with $\mathbf{S}_w + \lambda \Omega$ and then LDA proceeds as usual, where Ω is a symmetric and non-negative definite penalty matrix. The choice of Ω depends on the problem. If the data are log-spectra or images, Ω is defined in such a way so as to force nearby components of discriminant vectors to be similar. The main problem with RDA and PDA is that they do not scale well. In applications such as face recognition and cancer classification with gene expression profiling, the dimensionality of covariance matrices are often more than ten thousand. It is not practical for RDA and PDA to process such large covariance matrices, especially when the computing platform is made of PCs.

Recently, several methods that play with the null space of $\mathbf{S}_w$ have been widely investigated. A well-known null subspace method is the LDA+PCA method [Chen et al., 2000]. When $\mathbf{S}_w$ is of full rank, the LDA+PCA method just calculates the maximum eigenvectors of $\mathbf{S}_t^{-1}\mathbf{S}_b$ to form the transformation matrix. Otherwise, a two-stage procedure is employed. First, the data are transformed into the null space $\mathcal{V}_0$ of $\mathbf{S}_w$. Second, it tries to maximize the between-class scatter in $\mathcal{V}_0$, which is accomplished by performing PCA on the between-class scatter matrix in $\mathcal{V}_0$. Although this method solves the small sample size problem, it could be sub-optimal because it maximizes the between-class scatter in the null space of $\mathbf{S}_w$ instead of the original input space. For example, the performance of the LDA+PCA method drops significantly when $n - c$ is close to the dimensionality D, where n is the number of samples and c is the number of classes. The reason is that the dimensionality of the null space $\mathcal{V}_0$ is very small in this situation, and too much information is lost when we try to extract the discriminant vectors in $\mathcal{V}_0$. LDA+PCA also needs to calculate the rank of $\mathbf{S}_w$, which is an ill-defined operation due to floating-point imprecision. Another problem with LDA+PCA is that the computational complexity of determining the null space of $\mathbf{S}_w$ is very high. In [Huang et al., 2002], a more efficient null subspace method was proposed, which has the same accuracy as LDA+PCA. This method first removes the null space of $\mathbf{S}_t$, which has been proven to be the common null space of both $\mathbf{S}_b$ and $\mathbf{S}_w$, and useless for discrimination. Then, LDA+PCA is performed in the lower-dimensional projected space. Direct LDA is another null space method that discards the null space of $\mathbf{S}_b$ [Yu and Yang, 2001]. This is achieved by diagonalizing first $\mathbf{S}_b$ and then diagonalizing $\mathbf{S}_w$, which is in the reverse order of conventional simultaneous diagonalization procedure [Fukunaga, 1990]. In Direct LDA, one may also employ $\mathbf{S}_t$ instead of $\mathbf{S}_w$. In this way, Direct LDA is actually equivalent to PCA+LDA [Yu and Yang, 2001]. So, Direct LDA may be regarded as a "unified PCA+LDA" since there is no separate PCA step.

3.3 Generalized Linear Discriminant Analysis

In this section, we will first present some properties of scatter matrices and then develop the generalized linear discriminant analysis. We will also develop a fast algorithm of GLDA to deal with the high dimensionality of scatter matrices. Finally, we will show that LDA+PCA produce the same results as GLDA when $rank(\mathbf{S}_t) = rank(\mathbf{S}_b) + rank(\mathbf{S}_w)$.

3.3.1 Method

It is known that Fisher's criterion may also be written as

$$J(\mathbf{W}) = \mathrm{tr} \left(\frac{\mathbf{W}^T \mathbf{S}_b \mathbf{W}}{\mathbf{W}^T \mathbf{S}_t \mathbf{W}} \right) \qquad (3.5)$$

which is exactly equivalent to Equation (3.1) when $\mathbf{S}_w$ is nonsingular [Fukunaga, 1990]. However, there is an important difference between Equations (3.1) and (3.5) when $\mathbf{S}_t$ is singular (so is $\mathbf{S}_w$). It is well known that the null space of $\mathbf{S}_w$ contains the important discriminatory information. As pointed out in [Friedman, 1989], the discriminant function is heavily weighted by the smallest eigenvalues of $\mathbf{S}_w$ and the directions associated with their eigenvectors. When the small sample size problem occurs, these smallest eigenvalues are estimated to be 0. That is, the corresponding eigenvectors are in the null space of $\mathbf{S}_w$. Besides, it was also experimentally shown that the null space of $\mathbf{S}_w$ is crucial for discriminant analysis [Chen et al., 2000]. Unfortunately, we cannot take the derivative of Equation (3.1) in the null space of $\mathbf{S}_w$ to find the optimal solution because the vectors in the null space of $\mathbf{S}_w$ are the singular points of Equation (3.1). In contrast, the null space of $\mathbf{S}_t$ is a subspace of the null space of $\mathbf{S}_b$, which is useless for extracting the discriminatory information [Huang et al., 2002]. Thus, we can safely take the derivative of Equation (3.5) out of the null space of $\mathbf{S}_t$ in order to find the optimal solution. Therefore, we will use Equation (3.5) as the optimization criterion in the rest of this chapter.

In our proposed method and associated lemmas, the Moore-Penrose inverse is used, which is defined as:

Definition 3.1 (Moore-Penrose Inverse) *A matrix* $\mathbf{A}^+$ *satisfying the following conditions is unique and is called the Moore-Penrose inverse of* $\mathbf{A}$:

$$\mathbf{A}\mathbf{A}^+\mathbf{A} = \mathbf{A} \qquad\qquad \mathbf{A}^+\mathbf{A}\mathbf{A}^+ = \mathbf{A}^+$$
$$(\mathbf{A}^+\mathbf{A})^T = \mathbf{A}^+\mathbf{A} \qquad\qquad (\mathbf{A}\mathbf{A}^+)^T = \mathbf{A}\mathbf{A}^+$$

One may also define the matrix 1-inverse $\mathbf{A}^-$ by requiring only the first condition. However, such an inverse is not unique in general. Although most of our results also hold for the matrix 1-inverse, we confine ourselves to the Moore-Penrose inverse in the rest of chapter for uniquity.

Lemma 3.2 $\mathbf{S}_t\mathbf{S}_t^+(\mathbf{x} - \mathbf{m}) = \mathbf{x} - \mathbf{m}$

Proof. First, we can prove

$$E[(\mathbf{I} - \mathbf{S}_t\mathbf{S}_t^+)(\mathbf{x} - \mathbf{m})] = (\mathbf{I} - \mathbf{S}_t\mathbf{S}_t^+)E[\mathbf{x} - \mathbf{m}] = 0$$

and

$$\begin{aligned}
\mathrm{cov}((\mathbf{I} - \mathbf{S}_t\mathbf{S}_t^+)(\mathbf{x} - \mathbf{m})) &= (\mathbf{I} - \mathbf{S}_t\mathbf{S}_t^+)\mathrm{cov}(\mathbf{x} - \mathbf{m})(\mathbf{I} - \mathbf{S}_t\mathbf{S}_t^+)^T \\
&= (\mathbf{I} - \mathbf{S}_t\mathbf{S}_t^+)\mathbf{S}_t(\mathbf{I} - \mathbf{S}_t\mathbf{S}_t^+)^T \\
&= (\mathbf{S}_t - \mathbf{S}_t\mathbf{S}_t^+\mathbf{S}_t)(\mathbf{I} - \mathbf{S}_t\mathbf{S}_t^+)^T \\
&= (\mathbf{S}_t - \mathbf{S}_t)(\mathbf{I} - \mathbf{S}_t\mathbf{S}_t^+)^T = 0
\end{aligned}$$

So,

$$(\mathbf{I} - \mathbf{S}_t\mathbf{S}_t^+)(\mathbf{x} - \mathbf{m}) = 0$$

i.e.,

$$\mathbf{S}_t\mathbf{S}_t^+(\mathbf{x} - \mathbf{m}) = \mathbf{x} - \mathbf{m}$$

∎

Lemma 3.2 means that any centered sample $\mathbf{x} - \mathbf{m}$ is the eigenvector of $\mathbf{S}_t\mathbf{S}_t^+$ corresponding to eigenvalue 1. In this sense, it is natural that both $\mathbf{S}_b$ and $\mathbf{S}_w$ are constituted by the eigenvectors of $\mathbf{S}_t\mathbf{S}_t^+$ because the column vectors of both $\mathbf{S}_b$ and $\mathbf{S}_w$ are just the linear combinations of centered samples. These will be shown in Lemma 3.3 and 3.4 below.

Define

$$\begin{aligned}
\mathbf{M} &= [E[\sqrt{p_1}(\mathbf{x} - \mathbf{m})|\mathcal{C}_1], \ldots, E[\sqrt{p_c}(\mathbf{x} - \mathbf{m})|\mathcal{C}_c]] \\
&= [\sqrt{p_1}(\mathbf{m}_1 - \mathbf{m}), \ldots, \sqrt{p_c}(\mathbf{m}_c - \mathbf{m})]
\end{aligned}$$

where c is the number of classes; $\mathbf{m}_i$ and p_i are the mean vector and *a priori* probability of class i, respectively; and $\mathbf{m}$ is the overall mean vector. So, $\mathbf{S}_b = \mathbf{M}\mathbf{M}^T$.

48

Lemma 3.3 $\mathbf{S}_t\mathbf{S}_t^+\mathbf{S}_b = \mathbf{S}_b$

Proof. By Lemma 3.2, we have

$$\mathbf{S}_t\mathbf{S}_t^+(\mathbf{x} - \mathbf{m}) = \mathbf{x} - \mathbf{m}$$

Obviously,

$$\mathbf{S}_t\mathbf{S}_t^+\sqrt{p_i}(\mathbf{x} - \mathbf{m}) = \sqrt{p_i}(\mathbf{x} - \mathbf{m})$$
$$\mathbf{S}_t\mathbf{S}_t^+ E[\sqrt{p_i}(\mathbf{x} - \mathbf{m})|\mathcal{C}_i] = E[\sqrt{p_i}(\mathbf{x} - \mathbf{m})|\mathcal{C}_i]$$

where $i = 1, \ldots, c$. Thus,

$$\mathbf{S}_t\mathbf{S}_t^+\mathbf{M} = \mathbf{M}$$
$$\mathbf{S}_t\mathbf{S}_t^+\mathbf{M}\mathbf{M}^T = \mathbf{M}\mathbf{M}^T$$

i.e.,

$$\mathbf{S}_t\mathbf{S}_t^+\mathbf{S}_b = \mathbf{S}_b$$

■

Using Lemma 3.3, it is straightforward to prove a similar lemma for $\mathbf{S}_w$

Lemma 3.4 $\mathbf{S}_t\mathbf{S}_t^+\mathbf{S}_w = \mathbf{S}_w$

Proof.

$$\mathbf{S}_t\mathbf{S}_t^+\mathbf{S}_w = \mathbf{S}_t\mathbf{S}_t^+(\mathbf{S}_t - \mathbf{S}_b) = \mathbf{S}_t\mathbf{S}_t^+\mathbf{S}_t - \mathbf{S}_t\mathbf{S}_t^+ S_b = \mathbf{S}_t - \mathbf{S}_b = \mathbf{S}_w$$

■

From Lemmas 3.3 and 3.4, and the fact that $\mathbf{S}_b$, $\mathbf{S}_w$, and $\mathbf{S}_t$ are symmetric, it is easy to show $\mathbf{S}_b\mathbf{S}_t^+\mathbf{S}_t = \mathbf{S}_b$ and $\mathbf{S}_w\mathbf{S}_t^+\mathbf{S}_t = \mathbf{S}_w$.

The following theorem is the foundation of the GLDA method. That is, Fisher's criterion (Equation (3.5)) is maximized by the largest eigenvectors of $\mathbf{S}_t^+\mathbf{S}_b$. When $\mathbf{S}_w$ is nonsingular, $\mathbf{S}_t$ is also nonsingular and

$\mathbf{S}_t^+$ is equal to $\mathbf{S}_t^{-1}$. Thus, GLDA coincides with the conventional LDA when $\mathbf{S}_w$ is nonsingular. Note that our method is very different from the naive method that simply replaces the inverse matrix $\mathbf{S}_w^{-1}$ with the pseudo-inverse matrix $\mathbf{S}_w^+$, which has no rigorous mathematical support. In fact, we know that Equation (3.1) is not valid when $\mathbf{S}_w$ is singular. It is meaningless to just use the pseudo-inverse matrix $\mathbf{S}_w^+$ in this situation. On the other hand, we carefully analyze the properties of scatter matrices and if the null spaces of scatter matrices contain the discriminatory information. Our method is supported by the rigorous proofs (see below) and thus mathematically well-founded. Loosely speaking, GLDA can be regarded as a special case of PDA in the sense that $\Omega = \mathbf{S}_b$ and $\lambda = 1$. Note that Ω should be chosen intelligently in PDA so that $\mathbf{S}_t + \lambda\Omega$ is invertible [Hastie and Tibshirani, 1995]. However, $\mathbf{S}_t + \mathbf{S}_b$ is usually singular in GLDA. Different from the general PDA, GLDA has a closed-form solution. Later, we will also develop a fast algorithm of GLDA to efficiently handle the high dimensionality of data. In contrast, PDA encounters the computational difficulties when dealing with very high dimensional data.

Theorem 3.5 *Fisher's criterion (Equation (3.5)) is maximized by the largest eigenvectors of $\mathbf{S}_t^+\mathbf{S}_b$.*

Proof. Similar to the procedure for obtaining (3.4), we obtain the following equation by taking the derivative of (3.5) with respect to $\mathbf{W}$ and equating it to zero.

$$\mathbf{S}_b\mathbf{W} = \mathbf{S}_t\mathbf{W}(\mathbf{W}^T\mathbf{S}_t\mathbf{W})^{-1}(\mathbf{W}^T\mathbf{S}_b\mathbf{W}) \qquad (3.6)$$

Note that we have to restrict the domain of $\mathbf{W}$ to be outside the null space of $\mathbf{S}_t$ in order to take the derivative. As we mentioned before, the null space of $\mathbf{S}_t$ is a subspace of the null space of $\mathbf{S}_b$, which does not contain any discriminatory information [Huang et al., 2002]. Thus, such a restriction does not limit the discriminant capacity of the method.

Since the null space of $\mathbf{S}_t$ is a subspace of the null space of $\mathbf{S}_b$ [Huang et al., 2002], we can simultaneously diagonalize two symmetric matrices

50

$\mathbf{W}^T\mathbf{S}_b\mathbf{W}$ and $\mathbf{W}^T\mathbf{S}_t\mathbf{W}$ to $\mathbf{\Lambda}=\mathrm{diag}[\lambda_1,\dots,\lambda_d]$ and $\mathbf{I}$ [Fukunaga, 1990]:

$$\mathbf{P}^T(\mathbf{W}^T\mathbf{S}_b\mathbf{W})\mathbf{P} = \mathbf{\Lambda} \tag{3.7}$$

$$\mathbf{P}^T(\mathbf{W}^T\mathbf{S}_t\mathbf{W})\mathbf{P} = \mathbf{I} \tag{3.8}$$

where $\mathbf{P}$ is a $d \times d$ nonsingular matrix and d is less than or equal to the rank of $\mathbf{S}_t$. Besides, $\mathbf{\Lambda} \geq 0$ because $\mathbf{S}_b$ is positive semidefinite [Fukunaga, 1990]. So, we have

$$\mathbf{W}^T\mathbf{S}_b\mathbf{W} = (\mathbf{P}^{-1})^T\mathbf{\Lambda}\mathbf{P}^{-1} \tag{3.9}$$

$$\mathbf{W}^T\mathbf{S}_t\mathbf{W} = (\mathbf{P}^{-1})^T\mathbf{P}^{-1} \tag{3.10}$$

Using Equation (3.9) and (3.10), we can simplify the right hand side of Equation (3.6) as follows:

$$
\begin{aligned}
&\mathbf{S}_t\mathbf{W}(\mathbf{W}^T\mathbf{S}_t\mathbf{W})^{-1}(\mathbf{W}^T\mathbf{S}_b\mathbf{W}) \\
&= \mathbf{S}_t\mathbf{W}[(\mathbf{P}^{-1})^T\mathbf{P}^{-1}]^{-1}[(\mathbf{P}^{-1})^T\mathbf{\Lambda}\mathbf{P}^{-1}] \\
&= \mathbf{S}_t\mathbf{W}\mathbf{P}\mathbf{\Lambda}\mathbf{P}^{-1}
\end{aligned}
\tag{3.11}
$$

Then combining (3.6) and (3.11) leads to

$$\mathbf{S}_b\mathbf{W} = \mathbf{S}_t\mathbf{W}\mathbf{P}\mathbf{\Lambda}\mathbf{P}^{-1} \tag{3.12}$$

$$\mathbf{S}_b\mathbf{W}\mathbf{P} = \mathbf{S}_t\mathbf{W}\mathbf{P}\mathbf{\Lambda} \tag{3.13}$$

Plugging in $\mathbf{S}_b = \mathbf{S}_b\mathbf{S}_t^+\mathbf{S}_t$, we obtain

$$\mathbf{S}_b\mathbf{S}_t^+\mathbf{S}_t\mathbf{W}\mathbf{P} = \mathbf{S}_t\mathbf{W}\mathbf{P}\mathbf{\Lambda} \tag{3.14}$$

Denoting $\mathbf{K} = \mathbf{S}_t\mathbf{W}\mathbf{P}$, (3.14) can be expressed as

$$\mathbf{S}_b\mathbf{S}_t^+\mathbf{K} = \mathbf{K}\mathbf{\Lambda} \tag{3.15}$$

which means that the column vectors of $\mathbf{K}$ are the eigenvectors of $\mathbf{S}_b\mathbf{S}_t^+$ and $\lambda_i, i = 1,\dots,d$ are the corresponding eigenvalues.

Because Fisher's criterion is invariant under any nonsingular linear transformation [Fukunaga, 1990], we also have

$$
\begin{aligned}
J(\mathbf{W}) = J(\mathbf{W}\mathbf{P}) &= \mathrm{tr}((\mathbf{P}^T\mathbf{W}^T\mathbf{S}_t\mathbf{W}\mathbf{P})^{-1}(\mathbf{P}^T\mathbf{W}^T\mathbf{S}_b\mathbf{W}\mathbf{P})) \\
&= \mathrm{tr}(\mathbf{\Lambda}) = \lambda_1 + \cdots + \lambda_d
\end{aligned}
$$

51

Hence, in order to maximize the objective function, we have to choose the column vectors of $\mathbf{K}$ as the d largest eigenvectors of $\mathbf{S}_b\mathbf{S}_t^+$. Since $\mathbf{K} = \mathbf{S}_t\mathbf{W}\mathbf{P}$, we have

$$
\begin{aligned}
J(\mathbf{S}_t^+\mathbf{K}) &= J(\mathbf{S}_t^+\mathbf{S}_t\mathbf{W}\mathbf{P}) = J(\mathbf{S}_t^+\mathbf{S}_t\mathbf{W}) \\
&= \mathrm{tr}((\mathbf{W}^T\mathbf{S}_t\mathbf{S}_t^+\mathbf{S}_t\mathbf{S}_t^+\mathbf{S}_t\mathbf{W})^{-1}(\mathbf{W}^T\mathbf{S}_t\mathbf{S}_t^+\mathbf{S}_b\mathbf{S}_t^+\mathbf{S}_t\mathbf{W})) \\
&= \mathrm{tr}((\mathbf{W}^T\mathbf{S}_t\mathbf{W})^{-1}(\mathbf{W}^T\mathbf{S}_b\mathbf{W})) = J(\mathbf{W})
\end{aligned}
$$

because $\mathbf{S}_t\mathbf{S}_t^+\mathbf{S}_b = \mathbf{S}_b$ and $\mathbf{S}_t\mathbf{S}_t^+\mathbf{S}_t = \mathbf{S}_t$. In other words, the optimal transformation $\mathbf{W}$ is

$$
\mathbf{W} = \mathbf{S}_t^+\mathbf{K} \tag{3.16}
$$

By Equation (3.15), we have

$$
\mathbf{S}_t^+\mathbf{S}_b\mathbf{S}_t^+\mathbf{K} = \mathbf{S}_t^+\mathbf{K}\Lambda \tag{3.17}
$$

$$
\mathbf{S}_t^+\mathbf{S}_b\mathbf{W} = \mathbf{W}\Lambda \tag{3.18}
$$

i.e., the column vectors of $\mathbf{W}$ are the eigenvectors of $\mathbf{S}_t^+\mathbf{S}_b$. ∎

Since the rank of $\mathbf{S}_b$ is $c - 1$, there are only $c - 1$ eigenvectors to constitute the mapping matrix $\mathbf{W}$. In LDA/GLDA, we also require $\mathbf{W}$ to be orthonormal, which help preserve the shape of the distribution of the data. It is well-known that the mapping matrix $\mathbf{W}$ is not unique since $J(\mathbf{W}\mathbf{P}) = J(\mathbf{W})$. Since we require the mapping matrix to be orthonormal, $\mathbf{P}$ has to be orthonormal too. Thus, the effect of $\mathbf{P}$ is only to transform the discriminant space to another same-dimensional isometric space without any change to the geometric properties of patterns. That is, $\mathbf{P}$ has no influence on the classification and the importance of genes measured by $\mathbf{W}$. So, we can still rank marker genes by the absolute values of the elements of column vectors of $\mathbf{W}$.

3.3.2   A Fast Algorithm

Although the above GLDA solves the small sample size problem, we cannot apply it directly without developing a fast algorithm. Note that, both $\mathbf{S}_t$ and $\mathbf{S}_b$ have the dimensionality $D \times D$, where D is the dimensionality of the input space. Since D is usually many thousands, it is

52

very time and memory consuming to calculate $\mathbf{S}_t^+$ and the eigenvectors of $\mathbf{S}_t^+\mathbf{S}_b$ using common computational methods. In what follows, we will devise a fast algorithm for GLDA. First, we notice that $\mathbf{S}_t$ and $\mathbf{S}_t^+$ can be represented as $\boldsymbol{\Phi}\boldsymbol{\Theta}\boldsymbol{\Phi}^T$ and $\boldsymbol{\Phi}\boldsymbol{\Theta}^{-1}\boldsymbol{\Phi}^T$, respectively, where the columns of $\boldsymbol{\Phi}$ are the eigenvectors of $\mathbf{S}_t$ and the diagonal elements of the diagonal matrix $\boldsymbol{\Theta}$ are the corresponding eigenvalues. Besides, $\boldsymbol{\Phi}^T\boldsymbol{\Phi} = \mathbf{I}$. Thus, we can rewrite Equation (3.18) as

$$\mathbf{S}_t^+\mathbf{S}_b\boldsymbol{\Phi}\boldsymbol{\Theta}^{-\frac{1}{2}}(\boldsymbol{\Theta}^{\frac{1}{2}}\boldsymbol{\Phi}^T(\boldsymbol{\Phi}\boldsymbol{\Phi}^T)^{-1}\mathbf{W}) = \boldsymbol{\Phi}\boldsymbol{\Theta}^{-\frac{1}{2}}(\boldsymbol{\Theta}^{\frac{1}{2}}\boldsymbol{\Phi}^T(\boldsymbol{\Phi}\boldsymbol{\Phi}^T)^{-1}\mathbf{W})\boldsymbol{\Lambda}$$

$$\boldsymbol{\Phi}\boldsymbol{\Theta}^{-1}\boldsymbol{\Phi}^T\mathbf{S}_b\boldsymbol{\Phi}\boldsymbol{\Theta}^{-\frac{1}{2}}(\boldsymbol{\Theta}^{\frac{1}{2}}\boldsymbol{\Phi}^T(\boldsymbol{\Phi}\boldsymbol{\Phi}^T)^{-1}\mathbf{W}) = \boldsymbol{\Phi}\boldsymbol{\Theta}^{-\frac{1}{2}}(\boldsymbol{\Theta}^{\frac{1}{2}}\boldsymbol{\Phi}^T(\boldsymbol{\Phi}\boldsymbol{\Phi}^T)^{-1}\mathbf{W})\boldsymbol{\Lambda}$$

So,

$$\boldsymbol{\Phi}\boldsymbol{\Theta}^{-\frac{1}{2}}(\boldsymbol{\Theta}^{-\frac{1}{2}}\boldsymbol{\Phi}^T\mathbf{S}_b\boldsymbol{\Phi}\boldsymbol{\Theta}^{-\frac{1}{2}})(\boldsymbol{\Theta}^{\frac{1}{2}}\boldsymbol{\Phi}^T(\boldsymbol{\Phi}\boldsymbol{\Phi}^T)^{-1}\mathbf{W})$$
$$= \boldsymbol{\Phi}\boldsymbol{\Theta}^{-\frac{1}{2}}(\boldsymbol{\Theta}^{\frac{1}{2}}\boldsymbol{\Phi}^T(\boldsymbol{\Phi}\boldsymbol{\Phi}^T)^{-1}\mathbf{W})\boldsymbol{\Lambda}$$

Multiplying $\boldsymbol{\Theta}^{\frac{1}{2}}\boldsymbol{\Phi}^T$ on the left to both sides of the above equation, we get

$$(\boldsymbol{\Theta}^{-\frac{1}{2}}\boldsymbol{\Phi}^T\mathbf{S}_b\boldsymbol{\Phi}\boldsymbol{\Theta}^{-\frac{1}{2}})(\boldsymbol{\Theta}^{\frac{1}{2}}\boldsymbol{\Phi}^T(\boldsymbol{\Phi}\boldsymbol{\Phi}^T)^{-1}\mathbf{W}) = (\boldsymbol{\Theta}^{\frac{1}{2}}\boldsymbol{\Phi}^T(\boldsymbol{\Phi}\boldsymbol{\Phi}^T)^{-1}\mathbf{W})\boldsymbol{\Lambda} \quad (3.19)$$

i.e., $\boldsymbol{\Theta}^{\frac{1}{2}}\boldsymbol{\Phi}^T(\boldsymbol{\Phi}\boldsymbol{\Phi}^T)^{-1}\mathbf{W}$ is the eigenvector matrix of $\boldsymbol{\Theta}^{-\frac{1}{2}}\boldsymbol{\Phi}^T\mathbf{S}_b\boldsymbol{\Phi}\boldsymbol{\Theta}^{-\frac{1}{2}}$. Let $\boldsymbol{\Psi}$ be the eigenvector matrix of $\boldsymbol{\Theta}^{-\frac{1}{2}}\boldsymbol{\Phi}^T\mathbf{S}_b\boldsymbol{\Phi}\boldsymbol{\Theta}^{-\frac{1}{2}}$. Thus, we obtain $\mathbf{W} = \boldsymbol{\Phi}\boldsymbol{\Theta}^{-\frac{1}{2}}\boldsymbol{\Psi}$ since $\boldsymbol{\Phi}\boldsymbol{\Theta}^{-\frac{1}{2}}\boldsymbol{\Psi} = \boldsymbol{\Phi}\boldsymbol{\Theta}^{-\frac{1}{2}}\boldsymbol{\Theta}^{\frac{1}{2}}\boldsymbol{\Phi}^T(\boldsymbol{\Phi}\boldsymbol{\Phi}^T)^{-1}\mathbf{W} = \mathbf{W}$. As shown later, this two-step algorithm is efficient and helps reveal the connections among GLDA and LDA+PCA.

To obtain the eigenvector matrix $\boldsymbol{\Psi}$ of $\boldsymbol{\Theta}^{-\frac{1}{2}}\boldsymbol{\Phi}^T\mathbf{S}_b\boldsymbol{\Phi}\boldsymbol{\Theta}^{-\frac{1}{2}}$, we need first get the eigenvector and eigenvalue matrices $\boldsymbol{\Phi}$ and $\boldsymbol{\Theta}$ of $\mathbf{S}_t$, which can be efficiently obtained via SVD. Note that, $\mathbf{S}_t$ is estimated by $\frac{1}{n}\sum_{i=1}^{n}(\mathbf{x}_i - \mathbf{m})(\mathbf{x}_i - \mathbf{m})^T$ and can be expressed in the form $\mathbf{S}_t = \mathbf{X}\mathbf{X}^T$ with $\mathbf{X} = \frac{1}{\sqrt{n}}[(\mathbf{x}_1 - \mathbf{m}), \ldots, (\mathbf{x}_n - \mathbf{m})]$. Thus, we can obtain the eigenvalue matrix $\boldsymbol{\Theta}$ and the eigenvector matrix $\boldsymbol{\Phi}$ of $\mathbf{S}_t$ through the SVD of $\mathbf{X}$. Then we need obtain the eigenvector matrix $\boldsymbol{\Psi}$ of $\boldsymbol{\Theta}^{-\frac{1}{2}}\boldsymbol{\Phi}^T\mathbf{S}_b\boldsymbol{\Phi}\boldsymbol{\Theta}^{-\frac{1}{2}}$. Recall that $\mathbf{S}_b = \mathbf{M}\mathbf{M}^T$ with $\mathbf{M} = [\sqrt{p_1}(\mathbf{m}_1 - \mathbf{m}), \ldots, \sqrt{p_c}(\mathbf{m}_c - \mathbf{m})]$. Thus, we can obtain $\boldsymbol{\Psi}$ by the SVD of $\boldsymbol{\Theta}^{-\frac{1}{2}}\boldsymbol{\Phi}^T\mathbf{M}$. Finally, we obtain the mapping matrix $\mathbf{W} = \boldsymbol{\Phi}\boldsymbol{\Theta}^{-\frac{1}{2}}\boldsymbol{\Psi}$. A concise description of the algorithm is shown in Figure 3.1.

```
ALGORITHM GENERALIZED LINEAR DISCRIMINANT ANALYSIS
Input: A dataset of $n$ samples from $c$ classes with labels.
Output: The mapping matrix $\mathbf{W}$.
Method:
  1: Calculate $\mathbf{M} = [\sqrt{p_1}(\mathbf{m}_1 - \mathbf{m}), \ldots, \sqrt{p_c}(\mathbf{m}_c - \mathbf{m})]$ and $\mathbf{X} = \frac{1}{\sqrt{n}}[(\mathbf{x}_1 - \mathbf{m}), \ldots, (\mathbf{x}_n - \mathbf{m})]$.
  2: Perform the SVD $\mathbf{X} = \mathbf{\Phi}\mathbf{\Theta}^{\frac{1}{2}}\mathbf{U}^T$.
  3: Perform the SVD $\mathbf{\Theta}^{-\frac{1}{2}}\mathbf{\Phi}^T\mathbf{M} = \mathbf{\Psi}\mathbf{\Lambda}^{\frac{1}{2}}\mathbf{V}^T$.
  4: $\mathbf{W} = \mathbf{\Phi}\mathbf{\Theta}^{-\frac{1}{2}}\mathbf{\Psi}$
```

Figure 3.1: Fast Algorithm of Generalized Linear Discriminant Analysis.

3.3.3 Discussion

In this section, we will discuss the relation among GLDA and LDA+PCA. Particularly, we show that LDA+PCA produces the same results as GLDA when $rank(\mathbf{S}_t) = rank(\mathbf{S}_b) + rank(\mathbf{S}_w)$. When the condition does not hold, however, LDA+PCA could give the sub-optimal solutions as shown in the experiments.

Our discussion will be based on the LDA+PCA algorithm of Huang *et al.* Huang *et al.* improved the original LDA+PCA by removing the null space of $\mathbf{S}_t$ first [Huang et al., 2002]. The algorithm of Huang *et al.* is more efficient and has the same performance as the original LDA+PCA. The algorithm of Huang *et al.* has three steps: (i) remove the null space of $\mathbf{S}_t$ by the transformation of the eigenvector and eigenvalue matrix of $\mathbf{S}_t$; (ii) map the data into the null space of $\mathbf{S}_w$; (iii) remove the null space of $\mathbf{S}_b$. Compared with the algorithm of Huang *et al.*, our algorithm performs only the step (i) and (iii). In fact, the step (ii) is not necessary and can thus be skipped according to the following lemma. First, let the null space of a D-dimensional matrix $\mathbf{A}$ be denoted as $\text{Null}(\mathbf{A}) = \{\mathbf{x}|\mathbf{A}\mathbf{x} = 0, \mathbf{x} \in \mathcal{R}^D\}$, the range of $\mathbf{A}$ as $\text{Range}(\mathbf{A}) = \{\mathbf{A}\mathbf{x}|\mathbf{x} \in \mathcal{R}^D\}$, and $Dim(\cdot)$ be the dimensionality of a space.

Lemma 3.6 *If the condition $rank(\mathbf{S}_t) = rank(\mathbf{S}_b) + rank(\mathbf{S}_w)$ holds, then $Null(\mathbf{\Theta}^{-\frac{1}{2}}\mathbf{\Phi}^T\mathbf{S}_b\mathbf{\Phi}\mathbf{\Theta}^{-\frac{1}{2}}) \oplus Null(\mathbf{\Theta}^{-\frac{1}{2}}\mathbf{\Phi}^T\mathbf{S}_w\mathbf{\Phi}\mathbf{\Theta}^{-\frac{1}{2}}) = Range(\mathbf{\Phi}\mathbf{\Theta}^{-\frac{1}{2}}),$*

where Θ and Φ are the eigenvalue and eigenvector matrix of $\mathbf{S}_t$, respectively.

Proof. When $rank(\mathbf{S}_b) + rank(\mathbf{S}_w) = rank(\mathbf{S}_t)$, we also have

$$rank(\Theta^{-\frac{1}{2}}\Phi^T\mathbf{S}_b\Phi\Theta^{-\frac{1}{2}}) + rank(\Theta^{-\frac{1}{2}}\Phi^T\mathbf{S}_w\Phi\Theta^{-\frac{1}{2}})$$
$$= rank(\Theta^{-\frac{1}{2}}\Phi^T\mathbf{S}_t\Phi\Theta^{-\frac{1}{2}}) = rank(\Phi\Theta^{-\frac{1}{2}})$$

because $\Phi\Theta^{-\frac{1}{2}}$ has the full column rank and $rank(\Phi\Theta^{-\frac{1}{2}}) = rank(\mathbf{S}_t)$. Note that the null space $\text{Null}(\Theta^{-\frac{1}{2}}\Phi^T\mathbf{S}_b\Phi\Theta^{-\frac{1}{2}})$ has the dimensionality $rank(\Phi\Theta^{-\frac{1}{2}}) - rank(\Theta^{-\frac{1}{2}}\Phi^T\mathbf{S}_b\Phi\Theta^{-\frac{1}{2}})$. Similar result holds for $\mathbf{S}_w$. So, we have

$$Dim(\text{Null}(\Theta^{-\frac{1}{2}}\Phi^T\mathbf{S}_b\Phi\Theta^{-\frac{1}{2}})) + Dim(\text{Null}(\Theta^{-\frac{1}{2}}\Phi^T\mathbf{S}_w\Phi\Theta^{-\frac{1}{2}}))$$
$$= rank(\Phi\Theta^{-\frac{1}{2}}) \tag{3.20}$$

It is known that $\text{Null}(\mathbf{S}_t)$ is the intersection between $\text{Null}(\mathbf{S}_b)$ and $\text{Null}(\mathbf{S}_w)$ [Huang et al., 2002]. Thus, $\Theta^{-\frac{1}{2}}\Phi^T\mathbf{S}_b\Phi\Theta^{-\frac{1}{2}}$ and $\Theta^{-\frac{1}{2}}\Phi^T\mathbf{S}_w\Phi\Theta^{-\frac{1}{2}}$ have no common null space except $\{0\}$ because the mapping $\Phi\Theta^{-\frac{1}{2}}$ removes the null space of $\mathbf{S}_t$, i.e.

$$\text{Null}(\Theta^{-\frac{1}{2}}\Phi^T\mathbf{S}_b\Phi\Theta^{-\frac{1}{2}}) \cap \text{Null}(\Theta^{-\frac{1}{2}}\Phi^T\mathbf{S}_w\Phi\Theta^{-\frac{1}{2}}) = \{0\} \tag{3.21}$$

Based on Equations (3.20) and (3.21), we have

$$\text{Null}(\Theta^{-\frac{1}{2}}\Phi^T\mathbf{S}_b\Phi\Theta^{-\frac{1}{2}}) \oplus \text{Null}(\Theta^{-\frac{1}{2}}\Phi^T\mathbf{S}_w\Phi\Theta^{-\frac{1}{2}}) = \text{Range}(\Phi\Theta^{-\frac{1}{2}}) \tag{3.22}$$

∎

According to the above lemma, the null space of $\mathbf{S}_w$ and the range of $\mathbf{S}_b$ are complementary after removing the null space of $\mathbf{S}_t$. Thus, the data are already in the null space of $\mathbf{S}_w$ after they are mapped into the range of $\mathbf{S}_b$. Similarly, the data are already in the range of $\mathbf{S}_b$ after they are mapped into the null space of $\mathbf{S}_w$. Hence, we need only perform either the second step or the third step of Huang's algorithm when $rank(\mathbf{S}_t) = rank(\mathbf{S}_b) + rank(\mathbf{S}_w)$. So, our method will be the same as

Huang's algorithm in this situation. When the condition does not hold, however, our method is different from Huang's algorithm. In this case, the second step may cause too much information to be lost because the null space of $\mathbf{S}_w$ does not contain the whole column space of $\mathbf{S}_b$. Thus, LDA+PCA may have a very poor performance in this case as shown in the experimental results.

3.4 Experiments

In this section, we extensively compare our new methods with other methods in the literature on many public cancer datasets. First, we compare our methods with PDA [Hastie and Tibshirani, 1995, Hastie et al., 1994]. Then, we compare our methods with a typical univariate feature selection method [Dudoit et al., 2002]. This also serves to compare our methods with other widely used classifiers in cancer classification. We will also compare our methods with GA/KNN, which is a multivariate feature selection methods [Li et al., 2001a, Li et al., 2001b]. Finally, we compare our methods with support vector machine with recursive feature elimination (RFE) [Ramaswamy et al., 2001].

3.4.1 Data

In the experiments, we test our new method on seven public datasets, Leukemia [Golub et al., 1999], Colon [Alon et al., 1999], Prostate [Singh et al., 2002], Lymphoma [Alizadeh et al., 2000], SRBCT [Khan et al., 2001], Brain [Pomeroy et al., 2002], and GCM [Ramaswamy et al., 2001]. The leukemia dataset comprises 47 acute lymphoblastic leukemia (ALL, 38 B-cell ALL and 9 T-cell ALL) and 25 actue myeloid leukemia (AML) samples with the expression levels of 3571 genes. The colon dataset has the expression levels of 2000 filtered genes in 40 tumor and 22 normal colon tissues. The prostate dataset contains the expression levels of 6033 filtered genes in 52 tumor and 50 normal prostate tissues. The lymphoma dataset contains 62 samples of 3 classes: 42 diffuse large B-cell lymphoma, 9 observations of follicular lymphoma, and 11 cases of chronic lymphocytic leukemia. The expression levels of 4026 genes in these samples are used in the experiments. The SRBCT data consists of 63 samples of four subclasses of small, round blue cell tumors of childhood (SRBCTs), which include neuroblastoma (NB), rhabdomyosarcoma (RMS), non-Hodgkin lymphoma (NHL), and the Ewing family of tumors (EWS). The number of genes is 2308. The brain dataset is the dataset A in [Pomeroy et al., 2002] that has 42 samples of 10 medulloblastomas, 10 malignant gliomas, 10 atypical teratoid/rhabdoid tumors (AT/RTs), 8 primitive neuroectodermal tumors (PNETs) and 4 normal cerebella. The brain dataset contains the expression of 5597 genes. The GCM dataset is a very complicated dataset, which is a collection of 14 primary human cancer classes with the expression levels of 16063 genes [Ramaswamy et al., 2001]. These 14 common tumor classes account for $\approx$ 80% of new cancer diagnoses in the U.S. The GCM data consist of one training dataset of 144 primary tumor samples and one test dataset of 54 samples (46 primary and 8 metastatic). We combine the 144 samples of training dataset and 46 primary tumor samples of test dataset together and use the total 190 samples in the experiments. Before classification, the gene expression data are usually preprocessed in practice. In the experiments, we use the preprocessed data of the first six datasets by M. Dettling [Dettling, 2004], which can be downloaded from http://stat.ethz.ch/~dettling/bagboost.html. The preprocessing procedure includes base 10 log-transformation, normalization (with mean 0 and variance 1), and missing value imputation (by k-nearest neighbor) [Dettling, 2004, Dudoit et al., 2002]. Because the values in the GCM dataset are the raw average difference values (maybe negative) output from the Affymetrix software package, we do not perform the log-transformation and only normalize the values to mean 0 and variance 1. The properties of datasets are summarized in the top row of Table 3.1.

3.4.2 Results

The experimental procedure is as follows. For each dataset, we randomly split it into three parts in a class-proportional manner, of which two parts are used for training feature selection method and classifiers and the last part is kept for test. This procedure is repeated for 200 times and the

57

Table 3.1: Average classification error rates and standard deviations based on 200 runs.

	Features	Leukemia	Colon	Prostate	Lymphoma	SRBCT	Brain	GCM
c		2	2	2	3	4	5	14
n		72	62	102	62	63	42	190
D		3571	2000	6033	4026	2308	5597	16063
GLDA	$c-1$	3.1±2.8	14.5±5.7	7.6±3.7	0.05±0.47	1.9±2.6	16.6±8.8	17.9±3.8
PDA	$c-1$	3.3±2.7	14.0±5.7	N/A	0.17±0.88	1.7±2.4	N/A	N/A
RandFor	200	3.0±3.2	14.6±6.2	8.0±4.5	0.76±2.31	2.0±2.8	23.6±8.3	28.1±4.5
SVM	200	2.0±2.5	13.7±6.4	8.6±4.5	1.07±2.26	2.3±3.1	22.5±9.7	32.8±4.4
DLDA	200	2.8±2.9	14.4±6.6	15.5±7.7	1.71±2.57	2.1±2.7	24.0±9.7	31.6±5.0
kNN	200	3.7±3.1	18.2±6.4	11.2±4.6	1.12±2.24	0.9±1.9	23.0±8.3	44.1±4.6
RandFor	50	3.8±3.6	14.8±6.0	7.9±4.8	2.29±3.63	1.6±2.9	25.3±9.8	34.1±4.6
SVM	50	3.2±3.4	13.5±5.8	8.4±4.5	2.07±3.29	1.8±3.2	25.7±10.1	39.3±4.6
DLDA	50	2.8±2.9	13.6±6.4	12.3±6.5	3.86±3.92	1.8±3.6	25.8±10.5	38.2±5.2
kNN	50	4.2±3.4	19.3±7.2	12.6±4.8	2.12±3.46	1.8±2.6	26.6±9.6	46.8±5.8
RandFor	10	4.6±3.8	15.9±6.4	8.8±4.3	3.93±4.33	17.0±10.3	39.2±12.5	45.4±5.0
SVM	10	3.5±3.5	13.3±5.7	8.2±4.2	4.76±5.01	19.0±10.6	38.8±12.1	49.8±5.8
DLDA	10	3.2±3.5	12.9±6.0	11.0±5.6	7.76±5.85	17.4±10.1	37.3±14.0	49.6±6.0
kNN	10	3.6±3.5	18.6±6.8	11.0±4.7	5.00±5.16	21.6±12.0	43.9±12.9	54.0±4.7

[a] On the datasets Prostate, Brain, and GCM, PDA encountered some computational difficulties due to the limited memory (1GB on our machine). Thus, its results are not available here.

[b] The results of random forests (RandFor), SVM, DLDA, and kNN are based on the feature selection method of Dudoit et al. [Dudoit et al., 2002].

[c] In the table, c is the number of classes, n is the number of samples, and D is the number of genes.

averages and standard deviations of error rates are listed in Table 3.1. As shown in Table 3.1, GLDA performs well overall. To compare GLDA with other methods that try to solve the small sample size problem, we also perform PDA on the datasets. We observe that PDA and GLDA achieve similar results on Leukemia, Colon, Lymphoma, and SRBCT. However, PDA could not be applied on the Prostate, Brain, and GCM because the datasets have very high dimensionalities and thus the available memory (1GB on our machine) is not sufficient for PDA.

In what follows, we compare our methods with a univariate feature selection method. Because it was reported that various feature selection methods have the similar performance for cancer classification [Li et al., 2004], it is sufficient to compare our methods only with a typical feature selection method. In particular, we compare our methods with the univariate feature selection method of Dudoit *et al.* [Dudoit et al., 2002], which employs a metric similar to LDA:

$$\frac{BSS(j)}{WSS(j)} = \frac{\sum_i \sum_k I(y_i = k)(\bar{x}_{kj} - \bar{x}_{.j})^2}{\sum_i \sum_k I(y_i = k)(x_{ij} - \bar{x}_{kj})^2} \qquad (3.23)$$

where $\bar{x}_{.j}$ denotes the average expression level of gene j across all samples and $\bar{x}_{kj}$ denotes the average expression level of gene j across samples belonging to class k. The genes with the largest BSS/WSS ratios will be used for training. In the experiments, we try to select various numbers of top ranked genes. Here, we report the results with 10, 50, and 200 top ranked genes. We choose this feature selection method for comparison because it has a similar standpoint as LDA. However, it selects genes independently but LDA fully considers the correlation among genes. Thus, this provides a good opportunity to investigate if our methods improve the performance of classification compared with univariate feature selection. After feature selection, we use linear support vector machines (SVM) [Vapnik, 1998], random forests [Breiman, 2001], k-nearest neighbor (kNN, $k = 1$ here) [Fix and Hodges, 1951], and diagonal linear discriminant analysis (DLDA) [Dudoit et al., 2002] for classification. Thus, the experiments also serve as a comparison between our methods and the aforementioned classification methods. SVM and random forests are state-of-the-art machine learning methods and have proven very powerful in many applications. Although DLDA and kNN are very simple, many researcher have reported that they work very well for cancer classification. The weighted voting method of Golub *et al.* is actually a minor

Table 3.2: The error rates of GA/KNN given 10, 50, or 200 top ranked genes.

Datasets	Runs	10	50	200
Prostate	100	7.3 ± 3.5	8.1 ± 4.1	8.6 ± 4.4
SRBCT	200	3.0 ± 3.4	1.3 ± 2.3	1.4 ± 2.3
Brain	200	31.5 ± 9.9	22.1 ± 8.2	21.0 ± 8.2
GCM	25	55.6 ± 5.0	41.7 ± 5.0	36.3 ± 3.6

variant of DLDA [Dudoit et al., 2002], and thus we do not include it in the experiments.

With 200 top ranked genes determined by the method of Dudoit *et al.*, all aforementioned methods achieve similar results as GLDA on Leukemia, Colon, Prostate, Lymphoma, and SRBCT. However, GLDA achieves much better accuracies than its competitors on the Brain and GCM, which are very hard instances because of their large numbers of classes and genes. Compared with other methods, our methods reduce the error rate from about 22.5% to 16.6% on Brain and from about 28.1% to 17.9% on GCM. Besides, we observe that, one cannot meet the recommended sample per class ratio (i.e., 5 – 10) with 200 selected genes. Thus, the estimated error rates of the competing methods may be highly biased and the reliability of their results is low. In contrast, we can easily meet the recommended ratio after the dimension reduction by GLDA, which makes the estimated error rates more accurate and trustable. For a meaningful comparison (of trustable error rates), it is more suitable to compare the accuracy between GLDA and other classifiers with 10 top ranked genes. For such a comparison, GLDA is clearly better than all other methods on all datasets except for Colon. For Colon, we observe that all methods have roughly the same (high) error rates and the error rates do not change much with different numbers of genes. It has been reported that the colon dataset has a sample contamination problem [Li et al., 2001a], and may not be a suitable benchmark dataset.

Besides univariate feature selection, we would also like to compare our methods with the multivariate feature selection methods. Currently, two multivariate methods, gene-pair-ranking [Bø and Jonassen, 2002] and GA/KNN [Li et al., 2001a, Li et al., 2001b], have been proposed in the literature. The gene-pair-ranking method gives each pair of genes a score reflecting how well the pair in combination distinguishes two classes. Because the gene-pair-ranking method does not show a significant superiority to other methods and is hard to extend to deal with general gene subsets due to time complexity [Bø and Jonassen, 2002], we compare our methods only with GA/KNN below. Since most methods can achieve a high accuracy on Leukemia, and Lymphoma with only 10 top genes ranked by univariate methods and Colon has a sample contamination problem, we only show the comparison on the Prostate, SRBCT, Brain, and GCM datasets. The experimental procedure is the same as before and the results are summarized in Table 3.2. However, GA/KNN is very slow and needs a couple of weeks/months to complete 200 runs on Prostate/GCM datasets. As a result, we only perform GA/KNN 100 and 25 times on Prostate and GCM, respectively. As shown in Table 3.1 and 3.2, our methods are better than GA/KNN overall although GA/KNN performs better than univariate feature selection methods. Note that GA/KNN achieves worse results with more genes on the Prostate dataset, which is different from the trends on other datasets. The reason is not clear. In principle, one may improve the performance of GA/KNN by replacing k-nearest neighbor with some advanced classifier such as SVM. Unfortunately, it will increase the running time significantly since these (advanced) classifiers have a (much) higher time complexity than kNN, which has no training procedure. In fact, GA/KNN is already very slow even with kNN. For example, it takes more than five days to complete 200 runs on Brain on an Athlon MP 2800+ machine. Our methods, on the other hand, need only several minutes. Recall that, the user of GA/KNN also has to determine many parameters, such as chromosome length, the number of chromosomes, termination metric, etc. In contrast, GLDA does not need any parameters. Given the results in Table 3.1 and 3.2, we also observe that the simple methods such as GLDA and GA/KNN that multivariately find marker genes can achieve better results than sophisticated classification methods (e.g. random forests and SVM) that combined with a univariate feature selection method, especially when the number of selected features is small. It indicates that the choice of feature selection/extraction methods may be more important than the choice of classifiers for cancer classification.

In [Ramaswamy et al., 2001], Ramaswamy *et al.* proposed recursive feature elimination (RFE) that uses SVMs for both classification and

Table 3.3: Leave-one-out error rates of GLDA and RFE on GCM.

	RFE[a][b]			GLDA	
Genes per Classifier	kNN OVA	SVM OVA	Avg. Genes	kNN	GLDA
30	65.3%	70.8%	212	67.4%	78.5%
92	68.0%	72.2%	461	65.3%	81.9%
281	65.7%	73.4%	1132	66.0%	81.9%
1073	66.5%	74.1%	3972	66.7%	85.4%
3276	66.3%	74.7%	9821	67.4%	85.4%
6400	64.2%	75.5%	14512	67.4%	84.7%
All	N/A	78.0%	All	67.4%	84.7%

[a] The results of RFE were obtained by Ramaswamy *et al.* [Ramaswamy et al., 2001].

[b] The SVM and kNN employed in RFE use one-versus-all (OVA) coding scheme for multiclass classification.

feature selection. Recall that GLDA can also be used to select genes by treating the elements in the mapping matrix as the weights of genes. Here we would like to compare our methods with RFE. Like RFE, we first train GLDA on all features/genes. Then we select a subset of genes and train GLDA again on the selected genes. Ramaswamy *et al.* applied this method on the GCM data. Since this is a multiclass problem and SVM is a binary classifier, Ramaswamy *et al.* tried both one-versus-all (OVA) and all-pairs (AP) output coding schemes. We list their leave-one-out cross validation results of OVA on 144 training samples in Table 3.3. The results using AP are not listed here since they are worse than those using OVA. For the OVA coding scheme, one need train c binary SVMs, where c is the number of classes ($c = 14$ in the discussion below). Each SVM uses its own selected genes. Thus, the total number of selected genes is $s \times c$, where s is the number of genes per classifier and listed in the first column of Table 3.3. Of course, there may exist some overlap among the marker genes of different SVMs. Because our GLDA can solve multiclass problems directly, we use a different experimental setting. After training GLDA, we choose the top $s \times (c - 1)$ ranked genes because there are only $c - 1$ eigenvectors of $\mathbf{S}_t^+ \mathbf{S}_b$. Here, s is the same as that in RFE for reasonable comparison. Due to the overlap among top ranked genes of each eigenvector, the total number of genes is less than $s \times (c-1)$. The actual average numbers of selected genes are listed in the fourth column of Table 3.3. We observe that the overlap rate is very high. For instance, the overlap rate is $1 - 461/(92 \times 13) = 61.5\%$ for $s = 92$, which indicates that most eigenvectors give high weights to these 461 genes. Note that our methods select the genes in one step, not recursively. After the selection, we perform kNN and GLDA on the selected genes. Compared with RFE, GLDA performs clearly better as shown in Table 3.3. The accuracy 85.4% is the highest reported accuracy on GCM as far as we know. It is also interesting that the accuracy of SVM decreases as the number of marker genes decreases. Our methods, however, do not show such a pattern. When the number of genes decreases, GLDA may achieve a better accuracy because the noise might be reduced. Furthermore, the accuracy decreases when the number of genes becomes too small because a lot of information may be lost. Finally, we observe that kNN performs slightly better with our feature selection method than with RFE.

3.5 Conclusion

Gene expression profiling has great potential for accurate cancer diagnosis. It also brings machine learning researchers two challenges, the curse of dimensionality and the small sample size problem. In this chapter, we have presented two novel methods to solve these two problems. Our extensive experiments on seven public datasets demonstrate that the methods are able to classify tumors robustly with a high accuracy. Besides cancer classification, our work may also find applications in other areas where the small sample size problem and the curse of dimensionality arise such as image recognition and web document classification.

Chapter 4

Gene Expression Analysis with Minimum Entropy Clustering

4.1 Introduction

When the cell undergoes a specific biological process, different subsets of its genes are expressed in different stages of the process. The particular genes expressed at a given stage (*i.e.* under certain conditions) and their relative abundance are crucial to the cell's proper function. Technologies for generating high-density arrays of cDNAs and oligonucleotides enable us to simultaneously observe the expression levels of many thousands of genes on the transcription level during important biological processes [Schena et al., 1995, Velculescu et al., 1995, Lockhart et al., 1996]. Such a global view of thousands of functional genes changes the landscape of biological and biomedical research. Large amounts of gene expression data have been generated. Elucidating the patterns hidden in these gene expression data is a tremendous opportunity for functional genomics.

A preliminary and common methodology for analyzing gene expression data is the clustering technique. Clustering is the process of partitioning the input data into groups or *clusters* such that objects in the same cluster are more similar among themselves than to those in other clusters. Clustering has proved very useful for discovering important information from gene expression data. For example, clustering can help identify groups of genes that have similar expression patterns under various conditions or across different tissue samples [Eisen et al., 1998, Hughes et al., 2000]. Such co-expressed genes are typically involved in related functions. Clustering is also often the first step to discover regulatory elements in transcriptional regulatory networks [Alon et al., 1999, Tavazoie et al., 1999]. Co-expressed genes in the same cluster are probably involved in the same cellular process and strong expression pattern correlation between those genes indicates co-regulation.

A large number of clustering algorithms have been proposed [Jain and Dubes, 1988, Jain et al., 1999, Everitt et al., 2001, Han and Kamber, 2000, Kohonen, 2001]. Due to their simplicity, hierarchical clustering, k-means, and self-organizing maps (SOMs) are probably most popular methods for biologists to analyze gene expression data. However, it is well known that these methods are not effective in many situations and may not fit data well. For example, k-means and SOM require users to specify the number of clusters in advance. In many situations, however, it is difficult for biologists to know the exact number of clusters. Although hierarchical clustering does not need the number of clusters, the user still needs decide how many groups and where to cut the tree of clusters after the clustering. Besides, these methods perform well only on "clean" data without outliers. For instance, the k-means method is sensitive to outliers since a small number of such data can substantially influence the mean value. However, one person's noise is another person's signal. Outliers often contain important hidden information and provide clues for unknown knowledge. For example, a gene with abnormal expression levels may be related to some disease. Moreover, many clustering algorithms can find only convex clusters. But gene expression data have very complicated structure in general and may contain non-convex clusters.

Each of aforementioned challenges has been extensively studied by researchers from different fields. Although many sophisticated methods have been proposed, almost of them address only one or two chal-

66

lenges. Thus, biologists have to study many different algorithms, which results in a long learning curve. Although some density-based and grid-based clustering methods, such as DBSCAN [Ester et al., 1996], DEN-CLUE [Hinneburg and Keim, 1998], STING [Wang et al., 1997], and WaveCluster [Sheikholeslami et al., 1998], address all three challenges, these methods are of limited use for gene expression analysis due to various reasons, mainly because they cannot be applied to high-dimensional data. Moreover, many methods requires (and are sensitive to) several input parameters. To specify good parameters, biologists have to understand the complicated mathematical properties of algorithms and have a lot of experience. Since biologists do not have the same mathematical/algorithmic background as algorithm designers, it discourages biologists more. Therefore, simple clustering methods, like k-means and hierarchical clustering, are chosen in most research work on gene expression analysis.

In this chapter, we try to develop a simple clustering method that requires only one input parameter and address all aforementioned challenges with reasonably good performance. The proposed method is based on information theory. In particular, we try to minimize the conditional entropy of clusters given the observations. Many mathematical facts, such as Fano's inequality and the probability error of the nearest neighbor method, indicates that our clustering criterion may be a good clustering criterion. The input parameter, bandwidth of Parzen window, controls the resolution of clustering. Since most genes are multi-functional and the functions can be roughly organized in to hierarchical structure, such a multi-resolution property is very useful for gene expression analysis. We develop an efficient iterative algorithm to optimize our clustering criterion with a nonparametric approach to estimate *a posteriori* probabilities. The experimental results on synthetic data and real gene expression data show that the new clustering algorithm performs very well.

The rest of the chapter is organized as follows. In Section 4.2, we will review related work including those on determining the number of clusters, density-based clustering, grid-based clustering, and spectral clustering methods that can find clusters with arbitrary shape, and outlier detection methods. Section 4.3 introduces the minimum conditional entropy criterion for clustering. A brief review of entropy and structural

α-entropy is also included in this section. In Section 4.4, we follow the nonparametric approach to estimate *a posteriori* probabilities and propose an iterative algorithm to minimize the entropy. In Sections 4.3 and 4.4, we will also review previous entropy-based clustering methods in comparison with our method. Section 4.5 describes the experimental results on both synthetic data and real data. Section 4.6 concludes the chapter with some directions of future research.

4.2 Related Work

In this section, we will review previous work on determining the number of clusters, clustering methods that are capable of finding clusters with arbitrary shape, and outlier detection. Note that we will not review the work that is in the framework of biclustering (also referred as to subspace clustering in data mining community), e.g. CLIQUE [Agrawal et al., 1998], because we confine us to the conventional cluster analysis in this chapter. Besides, we will also not review the clustering methods for categorical data in general since they are useless for gene expression analysis. For previous entropy-based clustering methods, we will review them in comparison with our method in Sections 4.3 and 4.4.

A major challenge in cluster analysis is the estimation of the optimal number of clusters. In fact, there might be no definite or unique answer as to what value the number of clusters should take. There have been numerous methods proposed to estimate the suitable number of clusters [Milligan and Cooper, 1985, Jain and Dubes, 1988, Gordon, 1999]. Given partitions for each number of clusters $1 < k \leq M$ (or $1 \leq k \leq M$), many methods calculate some summary statistic, frequently a function of the within-cluster (and probably between-cluster) sums of squares (e.g. Calinski-Harabasz statistic [Calinski and Harabasz, 1974]), to determine the optimal number of clusters, where M is less than or equal to the number of observations in the dataset. Gap statistic, proposed in [Tibshirani et al., 2001], follows a different approach, which compares the change in some within-cluster dispersion measure to that expected under a reference null distribution. In the context of model-based cluster analysis, determining the number of clusters is usually referred as to model selection. Many model selection criteria/methods have been proposed, such as Akaike's information criterion (AIC) [Akaike, 1973], Bayesian inference criterion (BIC) [Schwarz, 1978], Cheeseman-Stutz criterion [Cheeseman and Stutz, 1995], minimum message length (MML) [Wallace and Boulton, 1968], minimum description length (MDL) [Rissanen, 1985], etc. There are some clustering methods that try to automatically inferring the number of clusters. For example, ISODATA, a variation of k-means, splits a cluster when its variance is above a pre-specified threshold and merges two clusters when the distance between their centroids is below another pre-specified threshold [BALL and HALL, 1965]. In this way, it is possible to obtain partitions with the suitable number of clusters, provided proper threshold values are specified. In [Zelnik-Manor and Perona, 2004], a technique was proposed to determine the number of clusters for spectral clustering by analyzing eigenvectors of affinity matrix. Interestingly, many density-based and grid-based clustering methods, e.g. DBSCAN [Ester et al., 1996], DENCLUE [Hinneburg and Keim, 1998], STING [Wang et al., 1997], and WaveCluster [Sheikholeslami et al., 1998], do not need the number of clusters in advance although they usually introduce several other input parameters. Recently, researchers from bioinformatics community also started to consider the problem of estimating the number of clusters. For example, Dudoit and Fridlyand proposed a prediction-based resampling method, Clest, for estimating the number of clusters based on the reproducibility or predictability of cluster assignments [Dudoit and Fridlyand, 2002].

It is well known that k-means and its variations can find only spherical clusters. Recently, various density-based, grid-based, and spectral clustering methods have been proposed to detect clusters with arbitrary shape. Density-based clustering algorithms, e.g. DBSCAN [Ester et al., 1996], OPTICS [Ankerst et al., 1999], and DENCLUE [Hinneburg and Keim, 1998], try to find clusters based on density of data points in a region. The key idea is that for each point, the neighborhood of a given radius has to contain at least a minimum number of points. Grid-based clustering algorithms, such as STING [Wang et al., 1997], and WaveCluster [Sheikholeslami et al., 1998], first quantize the clustering space into a finite number of cells (hyper-rectangles) and then perform the required operations on the quantized space. Cells that contain more than a certain number of points are treated as dense and the dense cells are connected to form clusters. Spectral clustering methods are another type of methods that can reliably find clusters with non-convex shapes [Weiss, 1999, Kannan et al., 2000, Ng et al., 2001, Zelnik-Manor and Perona, 2004]. Spectral clustering methods are based on the top eigenvectors of affinity matrix although different methods disagree on exactly which eigenvectors to use and how to derive clusters from them. It was also observed that there are some interesting similarities between spectral clustering methods and kernel PCA [Cristianini et al., 2001, Schölkopf et al., 1998]. When data are large, however, spectral clustering methods are not efficient due to the eigen decomposition on the huge affinity matrix, which limits their applications on gene expression analysis.

As an important task in statistics and knowledge discovery, outlier detection has also been extensively studied by many researchers. According to Hawkins, an outlier is an observation that deviates so much from other observations as to arouse suspicions that it was generated by a different mechanism [Hawkins, 1980]. Most of the existing work on outlier detection were conducted in the field of statistics [Barnett and Lewis, 1994, Hawkins, 1980, Ruts and Rousseeuw, 1996, Yamanishi et al., 2000]. Given (or having constructed) a probabilistic model, almost of statistical methods employ some hypothesis test to detect outliers. Although statistical methods are mathematically justified, most of them are designed for univariate or bivariate data. Besides, they are computational intensive when the underlying probability distributions are unknown. Many other techniques, such as Kolmogorov complexity [Arning et al., 1996], replicator neural networks [Hawkins et al., 2002], kernel technique [Petrovskiy, 2003], also have been applied to detect outliers. Recently, distance-based methods [Knorr et al., 2000, Ramaswamy et al., 2000, Bay and Schwabacher, 2003] and density-based methods [Breunig et al., 2000, Tang et al., 2002, Aggarwal and Yu, 2001] become popular. These methods does not make any assumptions about the distribution of data and the definitions of outliers have the intuitive geometric meaning. Although these methods have many attractive properties, they may not be sufficient in the context of clustering since these pure outlier detection methods do not consider the cluster structure in the data. For example, it is hard to distinguish between low dense clusters and outliers without the consideration of cluster structure. A few clustering algorithms, such as CLARANS [Ng and Han, 1994], BIRCH [Zhang et al., 1996], CURE [Guha et al., 1998], DBSCAN [Ester et al., 1996], DENCLUE [Hinneburg and Keim, 1998], STING [Wang et al., 1997], and

WaveCluster [Sheikholeslami et al., 1998], are designed to be able to handle outliers/exceptions.

We notice that some density-based and grid-based clustering methods, such as DBSCAN, DENCLUE, STING, WaveCluster, do not require the number of clusters, are able to detect arbitrary shaped clusters, and are not sensitive to noise. Although these methods address all three challenges, they are not suitable for gene expression analysis due to various reasons. First, most of them were designed for analyzing low-dimensional spatial database and often meet difficulties on high-dimensional datasets. For example, it is very hard to extend WaveCluster to process the data of more than two dimensions. It was also reported DBSCAN could not be run with data having more than 10 dimensions [Agrawal et al., 1998]. Note that even a small number of dimensions with uniform distribution can lower the density in the space enough so that no clusters are found by DBSCAN [Agrawal et al., 1998]. In fact, it is a common issue for most density-based and grid-based clustering methods. Second, density-based and grid-based methods may meet problems or require a large number of parameters when clusters have different density. Besides, these methods also suffers from the robustness problem as hierarchical clustering method. That is, when a dense string of points connecting two clusters, they could end up merging the two clusters. Thus, these methods are of limited use for gene expression analysis.

4.3   Minimum Entropy Clustering Criterion

In physics, entropy has important implications as the amount of "disorder" of a system. In information theory, the quantity entropy plays a central role as measures of information, choice and uncertainty. Mathematically, Shannon's entropy of a random variable X with a probability mass function $p(x)$ is defined as [Shannon, 1948]

$$H(X) = -\sum_{x} p(x) \log p(x) \tag{4.1}$$

Throughout the chapter, the base of log is 2. Entropy is the number of bits on the average required to describe a random variable. In fact, entropy is the minimum descriptive complexity of a random variable [Li and Vitányi, 1997].

Since entropy measures the amount of "disorder" of a system, it is natural to employ entropy as a clustering criterion. Let X be the random variable of observations taking values in $\mathbf{R}^d$ and C be the random variable of clusters taking values in $\{c_1, \ldots, c_m\}$ with probabilities $p_1, \ldots, p_m$, where d is the dimensionality of data and m is the number of clusters. Roughly, there are two types of minimum entropy clustering approaches in terms of which kind of conditional entropy between clusters and observations will be minimized. One tries to minimize $H(X|C)$ [Golchin and Paliwal, 1997, Barbara et al., 2002]. The other wants to minimize $H(C|X)$ [Roberts et al., 2000, Roberts et al., 2001, Lee and Choi, 2005]. The idea behind minimum $H(X|C)$ is that we hope that each cluster has a low entropy because data points in the same cluster should look similar. That is,

$$ J = H(X|C) = \sum_{j=1}^{m} p_j H(X|C = c_j) \tag{4.2} $$

where $H(X|C = c_j)$ is the entropy of cluster c_j, and p_j is the probability of cluster c_j such that $\sum_{j=1}^{m} p_j = 1$. Suppose each cluster c_j follows the d-dimensional Gaussian distribution with covariance matrix Σ_j. So, $H(X|c_j) = \log(2\pi e)^{d/2} + \frac{1}{2}\log|\Sigma_j|$ and

$$ J = \frac{1}{2} \sum_{j=1}^{m} p_j \log|\Sigma_j| \tag{4.3} $$

by discarding additive constant $\log(2\pi e)^{d/2}$. This falls into the conventional minimum variance clustering strategy. However, it does not guarantee one to find the correct cluster structure by minimizing variance. For example, we may achieve a small (total) variance if we divide a non-convex cluster into several smaller convex clusters. However, such a partition might not be meaningful.

Instead of minimizing $H(X|C)$, we prefer minimizing $H(C|X)$ due to the following reasons. In cluster analysis, we often assume that the data are drawn from a *mixed* source made up of a number of pure components that are each of homogeneous statistical structure. When partitioning data into groups, therefore, we want to minimize the uncertainty which cluster/source each datum is drawn from. In fact, the conditional entropy $H(C|X)$ just measures how uncertain we are of the random variable C of clusters on the average when we know X [Shannon, 1948]. First, let us look at

$$H(C|X = x) = -\sum_{j=1}^{m} p(c_j|x) \log p(c_j|x) \qquad (4.4)$$

In (4.4), we compute *a posteriori* probabilities $p(c_j|x)$ to determine how much information has been gained. $H(C|X = x)$ is maximized when all $p(c_j|x)$ are equal. In this case, the object x could come from any cluster equally probably, and thus we do not know which cluster the object x should belong to. This is also intuitively the most uncertain situation. On the other hand, $H(C|X = x)$ is minimized to 0 if and only if all the $P(c_j|x)$ but one are zero, this one having the value unity. That is, we are certain of the cluster of x only if $H(C|X = x)$ vanish. By integrating x on the whole data space, we obtain the total uncertainty of the partition that is just the conditional entropy $H(C|X)$

$$H(C|X) = -\int \sum_{j=1}^{m} p(c_j|x) \log p(c_j|x) p(x) dx \qquad (4.5)$$

Fano's inequality [Cover and Thomas, 1991] provides us another strong evidence that minimum $H(C|X)$ could be a good clustering criterion. Suppose we know a random variable X and we wish to guess the value of the correlated category information C. Fano's inequality relates the probability of error in guessing the random variable C to its conditional entropy $H(C|X)$. Suppose we employ a function $\hat{C} = f(X)$ to estimate C. Define the probability of error

$$P_e = \Pr\{\hat{C} \neq C\} \qquad (4.6)$$

Theorem 4.1 (Fano's Inequality)

$$H(P_e) + P_e \log(m - 1) \geq H(C|X) \qquad (4.7)$$

This inequality can be weakened to

$$1 + P_e \log m \geq H(C|X) \qquad (4.8)$$

Note that $P_e = 0$ implies that $H(C|X) = 0$. In fact, $H(C|X) = 0$ if and only if C is a function of X [Cover and Thomas, 1991]. That is, we can estimate C from X with zero probability of error if and only if C is a function of X. Fano's inequality indicates that we can estimate C with a low probability of error only if the conditional entropy $H(C|X)$ is small. In machine learning, P_e is expected to be small by the implicit assumption that X contains adequate information about C, i.e. $H(C|X)$ is small. Otherwise, it is meaningless to estimate C from X. Thus, the minimum $H(C|X)$ is a natural criterion for clustering.

So far, we have only considered Shannon's entropy. However, many measures of entropies have been introduced in the literature to generalize Shannon's entropy, e.g. Renyi's entropy [Renyi, 1961], Kapur's entropy [Kapur, 1967], and Havrda-Charvat's structural α-entropy [Havrda and Charvat, 1967], etc. We are particularly interested in the Havrda-Charvat's structural α-entropy here. The structural α-entropy is defined as

$$H^\alpha(X) = (2^{1-\alpha} - 1)^{-1} \left[\sum_x p^\alpha(x) - 1 \right] \tag{4.9}$$

where $\alpha > 0$ and $\alpha \neq 1$. With different degrees α, one can obtain different entropy measures. For example, when $\alpha \to 1$, we obtain Shannon's entropy:

$$\lim_{\alpha \to 1} H^\alpha(X) = -\sum_x p(x) \log p(x)$$

When $\alpha = 2$, we have the quadratic entropy:

$$H^2(X) = 1 - \sum_x p^2(x) \tag{4.10}$$

In the above equations, we discard the constant coefficients for simplicity. If the quadratic structural α-entropy is employed, we have

$$H^2(C|X) = 1 - \int \sum_{j=1}^m p^2(c_j|x) p(x) dx \tag{4.11}$$

Recall that, as a classification method, the nearest neighbor method has the following probability of error [Duda and Hart, 1973]:

$$R_{NN} = 1 - \int \sum_{j=1}^m p^2(c_j|x) p(x) dx \tag{4.12}$$

which is identical to Equation (4.11). Note that the probability of error R_{NN} is less than twice Bayes probability of error [Cover and Hart, 1967], which is the minimum probability of error over any other decision rule, based on the infinite sample set [Fukunaga, 1990]. It provides an evidence that the minimum $H^2(C|X)$ could potentially be a good criterion for clustering since Bayes probability of error (and the probability of error R_{NN}) is expected to be small by the implicit assumption that X contains adequate information about C as we mentioned before.

Another merit of structural α-entropy is that it satisfies the strong recursivity property. Suppose a random variable C has the distribution $P = (p_1, p_2, \ldots, p_m)$. In what follows, we write the entropy $H(X)$ as $H_m(p_1, p_2, \ldots, p_m)$. A measure of entropy $H_m(p_1, p_2, \ldots, p_m)$ will be said to have the recursivity property with respect to recursive function $g(p_1, p_2)$ if

$$H_m(p_1, p_2, \ldots, p_m)$$
$$= H_{m-1}(p_1 + p_2, \ldots, p_m) + g(p_1, p_2) H_2 \left(\frac{p_1}{p_1 + p_2}, \frac{p_2}{p_1 + p_2} \right)$$

holds for all $m \geq 3$ [Kapur, 1994]. Here $g(p_1, p_2)$ is a known continuous function ($g(p_1, p_2) = (p_1 + p_2)^\alpha$ for structural α-entropy). A stronger type of recursivity enables us to express H_m in term of two entropies, one of type H_{m-k+1} and the other of type H_k. In cluster analysis, the recursivity property is appreciated, especially when the data exhibit a nesting relationship between clusters. It is known that the only measures which satisfy the sum function property and the strong recursivity property are the Havrda-Charvat and Shannon's measures of entropy [Kapur, 1994].

In [Roberts et al., 2000, Roberts et al., 2001], Roberts *et al.* introduced a clustering method that wants to maximize the Kullback-Liebler divergence between the unconditional density of data and the conditional density of each cluster. Such an idea also results in the clustering criterion of minimizing the conditional entropy $H(C|X)$. [1] For each cluster, Roberts *et al.* use the EM algorithm to fit a (Gaussian) mixture

---

[1]In [Lee and Choi, 2005], Lee *et al.* follow the approach of Roberts *et al.* and replace Shannon's entropy with Renyi's quadratic entropy. With kernel density estimation (using Gaussian kernel), however, their method actually minimize $H(X|C)$ rather than $H(C|X)$ after replacing Kullback-Liebler divergence with quadratic distance measure.

model on the data. Then, the quasi-Newton method or reversible-jump Markov chain Monte Carlo is employed to (locally) minimize the conditional entropy. Although their methods show good performance on some low-dimensional datasets, it is not clear how the methods perform on high-dimensional data. Besides, it is also not clear if their methods are sensitive to outliers or not. Thus, their methods may not be sufficient for gene expression analysis.

4.4   The Clustering Algorithms

Given a dataset $\mathcal{X} = \{x_1, \ldots, x_n\}$, we want to partition data into groups that minimize the criterion

$$J = -\frac{1}{n} \sum_{i=1}^{n} \sum_{j=1}^{m} p(c_j|x_i) \ln p(c_j|x_i) \qquad (4.13)$$

Here, we just use Shannon's entropy to simplify our exposition. The proposed algorithms (to be discussed later) can be easily applied to the criterion with structural α-entropy.

4.4.1   Estimation of *a Posteriori* Probability

Before developing a clustering method to minimize $H(C|X)$, we have to estimate *a posteriori* probabilities $p(c_j|x)$. To estimate $p(c_j|x)$, we could employ some parametric method. However, it may not be appropriate since the underlying distribution of gene expression is unknown in general. An arbitrary choice of any particular distribution could lead to a very poor representation of the data. We therefore seek a nonparametric method for modelling the data. There are two kinds of nonparametric estimation techniques, *Parzen window density estimation* [Rosenblatt, 1956, Parzen, 1962] and *k-nearest neighbor density estimate* [Loftsgaarden and Quesenberry, 1965]. They are fundamentally very similar, but exhibit some different statistical properties. In what follows, we give a brief overview of these two nonparametric density estimation methods.

Consider estimating the value of a density function $p(x)$ at a point x. We may set up a small window $R(x)$ (e.g. hyper-cube or hyper-ellipsoid) around x. Then, the probability mass of $R(x)$ may be approximated by $p(x) \cdot v$ where v is the volume of $R(x)$. On the other hand, the probability mass of $R(x)$ may also be estimated by drawing a large number (say n) of samples from $p(x)$, counting the number (say k) of samples falling in $R(x)$, and computing k/n. Equating these two probabilities, we obtain an estimate of the density function as

$$p(x) = \frac{k}{n \cdot v} \qquad (4.14)$$

If we fix the volume v and let k be a function of x, we obtain Parzen window density estimate. On the other hand, we may fix k and let v be a function of x. More precisely, we extend the region $R(x)$ around x until the kth nearest neighbor is found. This approach is called the k-nearest neighbor density estimate.

By Bayes's rule, we have

$$p(c_j|x) = \frac{p(c_j)p(x|c_j)}{p(x)}$$

We may use n_j/n as the estimator of $p(c_j)$, where n_j is the number of points in cluster c_j. If Parzen window density estimate is employed, we have

$$p(c_j|x) = \frac{\dfrac{n_j}{n} \cdot \dfrac{k(x|c_j)}{n_j \cdot v}}{\dfrac{k(x)}{n \cdot v}} = \frac{k(x|c_j)}{k(x)} \qquad (4.15)$$

Thus, the estimate of $p(c_j|x)$ is just the ratio between the number of samples from cluster c_j and the number of all samples in the local region $R(x)$. If k-nearest neighbor estimate is used, we obtain

$$p(c_j|x) = \frac{\dfrac{n_j}{n} \cdot \dfrac{k}{n_j \cdot v(x|c_j)}}{\dfrac{k}{n \cdot v(x)}} = \frac{v(x)}{v(x|c_j)} \qquad (4.16)$$

There is an interesting comparison between the above estimation methods (Equations (4.15) and (4.16)) with the estimation methods in other minimum entropy clustering methods, such as those used in the method of Roberts *et al.* [Roberts et al., 2000, Roberts et al., 2001] and the method of Lee *et al.* [Lee and Choi, 2005]. Recall that Roberts *et al.* follow the parametric approach (mixture model) to estimate the density of each cluster while Lee *et al.* employ the kernel density estimation methods. Then both of them use Bayes's rule to estimate the *a posteriori* probability. In their estimation methods, global information, i.e. all data points, are used in the estimation of *a posteriori* probability since all data are used in the density estimation for each cluster. However, only local information, i.e. only neighbors of a point, is used to estimate the *a posteriori* probability associated with that point in our methods. Note that the goal is to partition data into groups rather than to estimating the density of each cluster. The latter task is very hard in general especially when the dimensionality of data is high because there are not sufficient observations practically due to the curse of dimensionality. Thus, the methods of Roberts *et al.* or Lee *et al.* may meet difficulties on high-dimensional data. In contrast, our methods avoid this problem. Note that our method is also different from the grid-based and density-based clustering methods, in which dense cells or points with dense neighborhood are used to form clusters. Compared with these methods, our methods focus on the ratios between conditional neighbor size and unconditional neighbor size without the consideration of the density of neighborhoods.

Because Parzen window method and k-nearest neighbor method are fundamentally very similar, we will employ only Parzen window estimate (4.15) in the rest of the chapter. In principle, one can easily replace Parzen window estimate (4.15) with k-nearest neighbor estimate (4.16) in our proposed algorithms (see later). However, it will potentially increase the time or space complexity of algorithm because a datum's k nearest neighbors of some cluster depend on the partition of data so that we have to recalculate the k-nearest neighbors whenever the partition changes or we need keep the distance matrix whose each row or column is sorted.

Parzen window method poses the problem of setting the parameter bandwidth of neighborhood. The bandwidth considerably affects the appearance of the density estimate and thus the partition of data. For example, if a very large (or infinite) bandwidth h_∞ is employed so that each datum has all observations as its neighbors, the *a posteriori* probability (4.15) will be just the *a priori* probability of each cluster. In this case, all data points have to been grouped into one and only one cluster to minimize the entropy to zero. If a very small (near to zero) bandwidth h_0 is used so that each datum has no neighbors except itself, on the other hand, the *a posteriori* probability (4.15) will be unity for the cluster of datum but zero for other clusters. In this case, each datum is in its own unique cluster and the entropy is already zero. In an analog of hierarchal clustering, h_∞ results in the partition at the highest level of hierarchy and h_0 produces the partition at the lowest level of hierarchy. With other bandwidths $h_0 < h < h_\infty$, it might partition data into groups at some level of hierarchy. That is, we may partition the data at different resolution levels (and thus different number of clusters) with different bandwidths. In this way, it actually provides a multi-resolution approach for cluster analysis by adjusting the bandwidth. Note that genes are usually multiple functional and can be organized in to hierarchical structure (more precisely, directed acyclic graphs). Besides, recall that there might be no definite or unique answer as to what value the number of clusters should take. Thus, such a multi-resolution approach is very much appreciated.

Bandwidth selection also has the important influence on the identification of outliers. Suppose outliers are defined as data points in small clusters (say less than five) that are far from major clusters, i.e. they are not neighbors of any points in the major clusters. Such a definition has an intuitive geometric meaning in the context of clustering. Based on this definition, there might be no outliers if the bandwidth is too large since every point is the neighbor of other points. If the bandwidth is near to zero, on the other hand, every point will be the outlier since every point is in its own unique cluster and is not the neighbor of other points. Simply put, bandwidth also determines the resolution of outlier identification.

Most work on bandwidth selection tries to select the "optimal" bandwidth in terms of density estimation error based on some criterion such as mean integrated squared error (MISE) [Turlach, 1993]. However, there are actually no (uniquely) "optimal" bandwidth for cluster analysis according to the above discussion. For cluster analysis, different bandwidths just result in partitions in different resolution levels. Suppose we want to determine the functions of some genes by clustering the expression patterns of genes, the choice of bandwidth obviously depends on what level of functions we are interested in. So, which resolution level is optimal depends on the research goals of analyzers and thus is a subjective concept. Therefore, it is not meaningful to select the so-called optimal bandwidth. Nevertheless, it is still useful to provide some "default" bandwidth for biologists to start the analysis. For multivariate data, the bandwidth is actually a symmetric positive definite matrix. For simplicity, we restrict the bandwidth matrix to a positive diagonal matrix $H = \text{diag}(h_1, \ldots, h_d)$, where d is the dimensionality of data. A variety of practical bandwidth selection methods, e.g. cross-validation, have been proposed for multivariate data [Chiu, 1996, Turlach, 1993, Scott, 1992]. For simplicity, we prefer the following bandwidth selection method [Scott, 1992]

$$h_i = \min(\hat{\sigma}_i, \frac{\hat{R}_i}{1.34}) \left(\frac{4}{(d+2)n} \right)^{1/(d+4)} \qquad (4.17)$$

where $\hat{\sigma}_i^2$ is the sample variance of ith variable, $\hat{R}_i$ is the interquartile range, and n is the sample size. Suppose the data are drawn from Gaussian distribution, such a selection is optimal in terms of asymptotic mean integrated squared error (AMISE) [Scott, 1992]. One drawback of this method is that the selected bandwidth is usually too small for high-dimensional data due to the curse of dimensionality so that most of data points have no neighbors except itself. Thus, we enlarge the bandwidth in practice with the following heuristics

$$\tilde{h}_i = \log(\log d + 1)h_i \qquad (4.18)$$

Although this method is very simple, it works fairly well for our purpose.

4.4.2 An Iterative Algorithm

In this section, we will develop a clustering algorithm to optimize

$$J = -\frac{1}{n} \sum_{i=1}^{n} \sum_{j=1}^{m} \frac{k(x_i|c_j)}{k(x_i)} \log \frac{k(x_i|c_j)}{k(x_i)} \qquad (4.19)$$

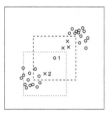

Figure 4.1: An illustration that the total entropy could increase when an object is assigned to the cluster containing most of its neighbors.

The proposed method is an iterative algorithm that reduces the entropy of an initial partition given by k-means. In principle, a (hill-climbing type) iterative algorithm starts with the system in some initial configuration. A standard rearrangement operation is applied to the system such that the objective function is improved. The rearrangement configuration then becomes the new configuration of the system, and the process is continued until no further improvement can be found. In our case, an intuitive idea to update the partition is to assign a data object x to the cluster containing most of its neighbors. This reassignment actually decreases the entropy associated with x because any change toward unequalization of the probabilities decreases the entropy [Shannon, 1948]. Suppose a point x is assigned to the cluster c_i currently and most neighbors of x are assigned to the cluster $c_j \neq c_i$. Moreover, suppose n_i neighbors of x belong to c_i and n_j neighbors of x belong to c_j such that $n_i < n_j$. After x is reassigned to cluster c_j, the difference between $n_i - 1$ and $n_j + 1$ is larger than that of n_i and n_j. So, we make the difference between the probabilities $p(c_i|x)$ and $p(c_j|x)$ larger, and thus reduce the entropy associated with x. Such an update, however, does not necessarily decrease the total entropy of the partition. Note that the entropy associated with the neighbors of x also changes after the reassignment. Figure 4.1 gives an illustration that the total entropy of the partition could increase after the reassignment. In Figure 4.1, the dashed line box represents the window $R(x)$ around data object 1, which has four neighbors belonging to cluster c_j denoted by 'x' and three neighbors (including object 1 itself) belonging to cluster c_i denoted by 'o'. The window around data object 2 is represented by the dotted line box. Note that all neighbors of object 2 belong to cluster c_i. If we reassign object 1 to cluster

ALGORITHM MINIMUM ENTROPY CLUSTERING ALGO-
RITHM
Input: A dataset containing n objects, the number of clusters
m, and an initial partition given by k-means.
Output: A set of at most m clusters that locally minimizes the
conditional entropy.
Method:

1: **repeat**
2:    **for** every object x in the dataset **do**
3:       **if** the cluster c_j containing most of the neighbors of x is
         different from the current cluster c_i of x **then**
4:          $h \leftarrow \sum_y (H'(C|y) - H(C|y))$
            {where y are neighbors of x, and x is also regarded
            as the neighbor of itself. $H(C|y)$ and $H'(C|y)$ are the
            entropy associated with y before and after assigning x
            to the cluster c_j, respectively.}
5:          **if** $h < 0$ **then**
6:             assign x to the cluster c_j
7:          **end if**
8:       **end if**
9:    **end for**
10: **until** no change

Figure 4.2: Minimum entropy clustering algorithm.

c_j, the entropy associated with object 2 will increase because it has a
neighbor in cluster c_j after the reassignment and thus the "disorder" of
the neighbors increases. Similarly, the entropy associated with the three
other objects in cluster c_j increases. So, the total entropy of partition
could increase although the entropy associated with object 1 decreases.
Based on this observation, we propose an algorithm that considers the
change of entropy associated with all neighbors of x, which is listed in
Figure 4.2.

Theorem 4.2 *Algorithm 4.2 converges after a sufficient number of iterations.*

82

Proof. Clearly, the total entropy of the partition decreases in every step. Thus, the algorithm 4.2 converges since the entropy is bounded by 0. ∎

Note that this algorithm could give a set of fewer than m clusters. The reason is that a cluster might migrate into another cluster to reduce the entropy during the iterations. This is different from most other clustering algorithms, which always return a given number of clusters. The speed of algorithm 4.2 depends on the number of iterations and the number of points that could be reassigned in each iteration, i.e. the points whose most neighbors belong to a different cluster. Usually, these kind of points exist in the overlap of clusters, and are only a small proportion of data. Besides, the number of neighbors could be regarded as constant in comparison with the whole data. So, the average time complexity of each iteration is usually less than $O(n)$ in practice. In the experiments, we found that the number of iterations is often very small, usually less than 20 for both synthetic and real data.

There is an interesting connection between our clustering algorithm and the regularized EM algorithm [Li et al., 2005]. When applied to finite mixture models, the regularized EM algorithm tries to maximize

$$\tilde{L}(\Theta; \mathcal{X}) = L(\Theta; \mathcal{X}) - \gamma H(C|X; \Theta)$$

where L is the likelihood function, γ is the regularization parameter, and Θ is the density parameter to estimate. Note that the regularizer $H(C|X; \Theta)$ is just the objective function of our clustering algorithm. Recall that there is the other regularization approach besides the above one [Poggio et al., 1985]. Simply put, among Θ that satisfy $L(\Theta; \mathcal{X}) \geq \bar{L}$, we want to find Θ that minimizes $H(C|X; \Theta)$, where $\bar{L}$ is a threshold. Clearly, our clustering algorithm can be loosely regarded as a special regularized EM algorithm in the latter regularization form. There are two important differences between the regularized EM algorithm and our clustering method. First, we employ k-means instead of the EM algorithm to find an initiation partition since we follow a nonparametric approach for clustering. In fact, k-means can be regarded as a special case of the EM algorithm [Bishop, 1995]. Besides, k-means is more efficient and more robust than the general EM algorithm on high-dimensional data. Second, when we minimize the regularizer (i.e. conditional entropy), we do not attempt to keep the likelihood of the partition larger than some threshold. Otherwise, we may not be able to find nonconvex clusters because the initial partition given by k-means consists of only spherical clusters.

4.5 Experiments

We have tested our minimum entropy clustering (MEC) algorithm on both synthetic data and real gene expression data, in comparison with k-means/medians, hierarchical clustering, SOM, and EM, which are widely used for clustering gene expression data. For k-means/medians, hierarchical clustering, and SOM, we used Eisen's implementations [Eisen et al., 1998]. For the EM algorithm, we used Fraley and Raftery's implementation [Fraley and Raftery, 2002]. To assess the quality of our algorithm, we need some objective external criteria. The external criteria could be the true class information, gene functional categories, *etc.* In order to compare clustering results against an objective external criterion, we employ adjusted Rand index [Hubert and Arabie, 1985] as the measure of agreement. Rand index [Rand, 1971] is defined as the number of pairs of objects that are either in the same group or in different groups in both partitions divided by the total number of pairs of objects. The Rand index lies between 0 and 1. When two partitions agree perfectly, the Rand index achieves the maximum value 1. A problem with Rand index is that the expected value of the Rand index between two random partitions is not a constant. This problem is corrected by the adjusted Rand index that assumes the generalized hypergeometric distribution as the model of randomness. The adjusted Rand index has the maximum value 1, and its expected value is 0 in the case of random clusters. A larger adjusted Rand index means a higher agreement between two partitions. The adjusted Rand index is recommended for measuring agreement even when the partitions compared have different numbers of clusters [Milligan and Cooper, 1986]. In what follows, the reported adjusted Rand index is averaged on 100 repeated experiments to reduce the influence of random initial partitions.

Table 4.1: Adjusted Rand index on the synthetic Gaussian data.

m	k-means	EM [a]	MEC [b]
2	0.535	0.802	0.704
3	0.452	0.620	0.610
4	0.349	0.386	0.384
5	0.280	0.294	0.448
6	0.222	0.229	0.542
7	0.190	0.195	0.633
8	0.163	0.168	0.593
9	0.147	0.163	0.526
10	0.134	0.153	0.502

[a] The EM algorithm uses the Gaussian mixture model.
[b] The MEC algorithm uses the structural α-entropy with $\alpha = 2..$

4.5.1 Synthetic Data

To give a visual illustration of our new method, we generate a two-dimensional synthetic data that consists of two highly-overlapped components following the Gaussian distribution as shown in Figure 4.3(a). In this experiment, we compare our MEC algorithm and the EM algorithm, both of which use the output of k-means as the initial partition. Table 4.1 lists the adjusted Rand index achieved by k-means, EM and MEC algorithms. Both EM and MEC significantly improve the initial partitions given by k-means. It is not surprising that the EM algorithm obtains the best results when the specified number of clusters is correct since its model perfectly matches the data. However, we often do not know the exact number of clusters in practice. When the specified number of clusters are not correct, both k-means and EM algorithm have very low adjusted Rand indexes. In contrast, the MEC algorithm still performs very well. Figure 4.3 gives a graphical representation of the clustering results when 8 clusters are specified. Other configurations give similar results. Both k-means and EM return 8 clusters as specified and neither represent the structure of the data correctly. The MEC algorithm, however, returns 5 clusters and gives the correct structure of the data. More precisely, two of the five clusters contain most of the data objects and represent the

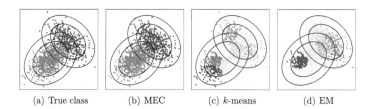

(a) True class   (b) MEC   (c) k-means   (d) EM

Figure 4.3: The synthetic two-component Gaussian data. Subfigure (a) represents the true class information. The left bottom component consists of 800 points. The upper right component consists of 400 points. The inner and outer contours are 2σ contour and 3σ contour, respectively. Subfigures (b), (c), and (d) are the clustering results when the specified number of clusters is 8. Different colors denote differnt clusters.

structure of the data well. The remaining three clusters consist of 5, 2, and 1 objects (denoted by '×'), respectively. These data objects stand apart from the bulk of the data. More precisely, they lie outside of the 3σ contour (*i.e.* the outer contour). Thus, these inconsistent data objects could be regarded as outliers. Many clustering algorithms are sensitive to the influence of outliers and can only perform well on "clean" data [Jain and Dubes, 1988, Everitt et al., 2001]. However, outliers often contain important hidden information and provide clues for unknown knowledge. For example, a gene with abnormal expression may be related to some disease. Figure 4.3(b) shows that the MEC algorithm can effectively identify outliers and correctly discover the structure of the main data simultaneously.

In the above synthetic Gaussian dataset, the clusters have convex shapes. However, many clusters do not possess a convex shape in practice. In what follows, we test our method in comparison with hierarchial clustering method, k-means, and SOM on a synthetic nonconvex dataset. For k-means, hierarchical clustering, and SOM, Euclidean distance is used to measure (dis)similarity. For hierarchical clustering, we employ the complete link algorithm. For MEC, we use Shannon entropy. The dataset contains two clusters generated by sine and cosine function with a white noise of standard deviation 0.1. Each cluster consists of 150 points. The data is shown in Figure 4.4(a), which is also the clus-

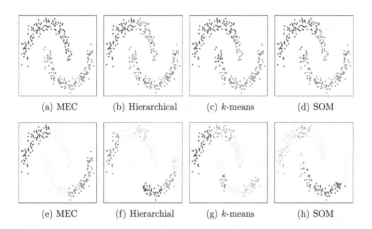

(a) MEC   (b) Hierarchical   (c) k-means   (d) SOM

(e) MEC   (f) Hierarchial   (g) k-means   (h) SOM

Figure 4.4: The synthetic nonconvex data. The dataset contains two clusters generated by sine and cosine functions with a Gaussian noise of mean 0 and standard deviation 0.1. Each cluster consists of 150 points. The first and second rows are the clustering results when the specified number of clusters are 2 and 8, respectively. For k-means, hierarchical clustering, and SOM, Euclidean distance is used to measure (dis)similarity. For hierarchical clustering, we employ the complete link algorithm. For MEC, we use Shannon entropy.

tering result of MEC when the specified number of clusters is 2. That is, MEC can perfectly cluster the dataset in this case. In contrast, all other three methods make some errors. When the specified number of clusters are not correct, our method can still return meaningful results. The second row of Figure 4.4 is the results when the specified number of clusters is 8. Similar results were obtained with other specified number of clusters. As shown in Figure 4.4(e), MEC returns four clusters in this case. Although it does not match the true clusters, it is still meaningful in practice. Note that the two reported clusters in each true cluster have very different patterns: one is up, the other is down. Thus, it is natural for us to interpret the dataset as four clusters.

4.5.2 Real Gene Expression Data

We used two gene expression data to test the MEC algorithm. The first data is the yeast galactose data on 20 experiments [Ideker et al., 2001]. For each cDNA array experiment, four replicate hybridizations were performed. Yeung *et al.* extracted a subset of 205 genes that are reproducibly measured, whose expression patterns reflect four functional categories in the Gene Ontology (GO) listings [Yeung et al., 2003]. The dataset contains approximately 8% of missing data. Yeung *et al.* applied k-nearest neighbor ($k = 12$) to impute all the missing values. We used this subset of data in our test with the four functional categories as the external knowledge. [2] For each cDNA array experiment, we used the average expression levels of four measurements for cluster analysis. Before clustering, we normalized the data for each gene to have mean 0 and variance 1 across experiments.

The experimental results are listed in Tables 4.2. Clearly, the MEC algorithm performs much better than k-means/medians, hierarchical clustering, SOM, and EM, especially when the specified number of clusters are far from the true number of clusters. Note that the EM algorithm

---

[2]We choose this subset rather than the whole dataset for experiments because it is very hard to determine the consistent functional categories for several thousand genes since many genes are multifunctional. Besides, it is also not easy to choose a suitable level of functional category for a large number of genes.

Table 4.2: Adjusted Rand index on the yeast galactose data.

Method	Setting	Specified number of clusters						
		4	5	6	7	8	9	10
MEC	Shannon's entropy	0.915	0.914	0.918	0.923	0.915	0.918	0.885
MEC	$\alpha = 2$	0.918	0.926	0.928	0.932	0.930	0.933	0.888
MEC	$\alpha = 3$	0.919	0.926	0.929	0.928	0.932	0.932	0.888
EM [a]		0.788	0.716	0.661	0.629	0.593	0.557	0.547
k-means	Euclidean dist.	0.806	0.746	0.671	0.628	0.547	0.534	0.494
k-means	Pearson corr.	0.806	0.739	0.667	0.604	0.538	0.511	0.454
k-median	Euclidean dist.	0.823	0.748	0.648	0.618	0.552	0.520	0.470
k-median	Pearson corr.	0.756	0.672	0.608	0.576	0.513	0.461	0.410
SOM	Euclidean dist.	0.845	0.853	0.674	0.556	0.443	0.382	0.342
SOM	Pearson corr.	0.825	0.845	0.675	0.559	0.446	0.382	0.348
Hierarchical[b]	Euclidean dist.	0.677	0.605	0.703	0.700	0.708	0.711	0.694
Hierarchical[b]	Pearson corr.	0.677	0.605	0.703	0.700	0.708	0.711	0.694

[a] In the EM algorithm, we assume that clusters have the diagonal covariance matrices, but with varying volume and shape. For more general ellipsoidal setting, the EM algorithm meets the computational problem (singular covariance matrix).

[b] For hierarchical clustering, we employ the complete link algorithm.

even gives worse partitions than k-means in some cases. One possible reason is that such a small size data is not sufficient for EM to estimate its parameters, which are a lot more than those of k-means. Another reason is that the data may follow some unknown distribution so that the assumption of Gaussian mixture is not suitable. For yeast galactose data, the MEC algorithm achieves a very high adjusted Rand index (> 0.9). This indicates that the MEC algorithm is capable of effectively grouping genes in the same functional category according to their expression levels. Moreover, the MEC algorithm achieves a higher adjusted Rand index even when the specified number of clusters is larger than the correct number (*i.e.* 4 in this case). The reason is that, when the specified number of clusters is larger than the correct one, the MEC algorithm could use the "extra" clusters to identify outliers and thus improve the quality of the final partition. In this yeast galactose data, the MEC algorithm identified two genes as outliers. They are YLR316C and STO1, belonging to functional category 3 (nucleobase, nucleoside, nucleotide and nucleic acid metabolism). In the original dataset, STO1 is incorrectly classified as functional category 2 (energy pathways, carbohydrate metabolism, and catabolism). To verify that these two genes are really outliers in this microarray experiment, we calculate the Pearson's correlation coefficients between them and the means of clusters. The correlation and coefficients and corresponding P-values are listed in Table 4.3. Since Pearson's correlation coefficient follows a t-distribution of $n - 2$ degrees of freedom (n is the number of arrays, e.g. $n = 20$ in this case), the statistical significance of a given correlation coefficient r is tested using a t-test with the hypotheses

$$H_0 : r = 0$$
$$H_1 : r \neq 0$$

A low P-value for this test (say, less than 0.05) indicates that there is evidence to reject the null hypothesis. That is, there is a statistically significant relationship between the two variables. Since all correlation coefficients have very large P-values in our tests, it indicates that these two gene have very different expression patterns from the others. Thus, they can be regarded as outliers in this experiment. Note that the outliers are relative to the dataset.

The above dataset is a typical gene expression dataset in which the expression levels of genes are measured at many time points or under

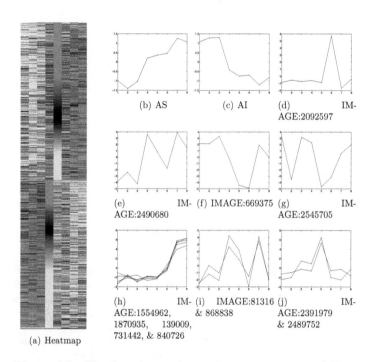

(a) Heatmap

(b) AS

(c) AI

(d)   IM-AGE:2092597

(e)   IM-AGE:2490680

(f) IMAGE:669375

(g)   IM-AGE:2545705

(h)   IM-AGE:1554962, 1870935,   139009, 731442, & 840726

(i)   IMAGE:81316 & 868838

(j)   IM-AGE:2391979 & 2489752

Figure 4.5: The clustering results on the prostate cancer cell lines dataset. In subfigure (a), each row represents a gene. The columns represent cell lines DU145, PPC-1, PC-3 (3 AI), 22Rv1, LAPC-4, LNCaP, MDA PCa 2a, and MDA PCa 2b (5 AS) from left to right. Subfigures (b) and (c) are the average expression levels of the top cluster (up-regulated across AS) and the bottom cluster (up-regulated across AI), respectively. Subfigures (d)–(j) are the expression levels of seven outlier groups, respectively. For subfigures (b)–(j), the horizontal axis is the cell lines in the same order as subfigure (a).

91

Table 4.3: The correlation coefficients between YLR316C, STO1 and the means of four clusters in the yeast galactose dataset.

Gene	Cluster	Correlation Coefficient	P-value
YLR316C	1	0.0538	0.8218
	2	0.1310	0.5819
	3	0.0608	0.7990
	4	0.2067	0.3818
STO1	1	0.0241	0.9198
	2	-0.0682	0.7750
	3	0.1117	0.6390
	4	-0.0913	0.7018

different conditions to elucidate genetic networks or some important biological process. Another type of gene expression data evaluates each gene in a single environment but in different types of tissues, for example body atlas data or tumor data. In what follows, we will test our method on a prostate cancer gene expression data. Since the goal is to characterize gene expression in cancer cell lines rather than to elucidate some biological process, it is not suitable to use the functional categories of genes as external information. If we know if a gene is related to some specific tumor (or its subtypes), we may use this information to calculate adjusted Rand index. However, only few tumor-specific molecular markers are currently known by molecular oncology. Thus, we will not compute the adjusted Rand index on the tumor gene expression dataset when evaluating the method. In [Zhao et al., 2005], Zhao *et al.* examined the gene expression profiles in androgen sensitive (AS) and androgen insensitive (AI) prostate cancer cell lines on a genome-wide scale. They measured the transcript levels of 27365 genes in five AS (LNCaP, LAPC-4, MDA PCa 2a, MDA PCa 2b, and 22Rv1) and three AI (PC-3, PPC-1, and DU145) prostate cell lines using cDNA microarrays. In particular, they selected 1703 clones representing 1261 unique genes that varied by at least 3-fold from the mean abundance in at least two cell lines. We perform our method on the expression levels of these 1703 clones. Before clustering, we impute the missing values using the k-nearest neighbor method ($k = 10$). Figure 4.5(a) is the clustering results when the specified number of clusters is 20. Other configurations have the similar results. As shown in Figure 4.5(a), there are two big clusters in the dataset. For the top cluster (872 genes) in the figure, genes are overall up-regulated across AS cells but down-regulated across AI cells. On the other hand, 818 genes in the bottom cluster are overall down-regulated across AS cells but up-regulated across AI cells. However, we observe that genes in neither clusters follow a uniform pattern across cell lines 22Rv1 and LAPC-4. According to Zhao (private communication), both 22Rv1 and LAPC-4 were established from Xenograpfts. Human cancer cells were implanted into immune deficient mice, and tumors grown in mice were taken out and implanted into new mice again. After serial passages, the cells became immortal. Maybe events happened during this process make them different from other cell lines that were established not as Xenograpfts.

Our method also found 7 small clusters containing 1, 2, or 5 genes, which are listed in Table 4.4. These small clusters are plotted in Figures 4.5(d) – 4.5(j). As a comparison, we also plot the average expression levels of the two big clusters in Figure 4.5(b) and 4.5(c), respectively. Clearly, these genes have either different patterns from those of the two big clusters or have very large (or very small) expression levels on some cell lines. Thus, they could be regarded as outliers.

4.6   Conclusion

From an information-theoretic point of view, we propose the minimum entropy clustering method for gene expression analysis. With a nonparametric approach for estimating *a posteriori* probabilities, an efficient iterative algorithm is established to minimize the entropy. The experimental results show that our new method performs well on various datasets. It would be interesting to extend this new algorithm to a subspace clustering method for finding clusters with different subsets of attributes (*i.e.* the bi-clustering problem). Besides, the presented idea could also be applied to supervised learning, although this chapter only focuses on unsupervised learning.

Table 4.4: The detected outliers in the prostate cancer cell lines dataset.

ID	Description
IMAGE:1554962	Homo sapiens transcribed sequence with weak similarity to protein ref:NP_060265.1 (H.sapiens) hypothetical protein FLJ20378 [Homo sapiens]; Hs.270149
IMAGE:1870935	FN1; fibronectin 1; Hs.418138
IMAGE:139009	FN1; fibronectin 1; Hs.418138
IMAGE:731442	COL4A3BP; collagen, type IV, alpha 3 (Goodpasture antigen) binding protein; Hs.21276
IMAGE:840726	Homo sapiens transcribed sequence with weak similarity to protein pir:S57447 (H.sapiens) S57447 HPBRII-7 protein - human; Hs.47026
IMAGE:81316	ARG99; ARG99 protein; Hs.401954
IMAGE:868838	HPGD; hydroxyprostaglandin dehydrogenase 15-(NAD); Hs.77348
IMAGE:2391979	DHRS2; dehydrogenase/reductase (SDR family) member 2; Hs.272499
IMAGE:2489752	PIP; prolactin-induced protein; Hs.99949
IMAGE:2092597	GAGE5; G antigen 5; Hs.278606
IMAGE:2490680	TRGV9; T cell receptor gamma variable 9; Hs.407442
IMAGE:669375	DKK1; dickkopf homolog 1 (Xenopus laevis); Hs.40499
IMAGE:2545705	CNN3; calponin 3, acidic; Hs.194662

Chapter 5

A General Framework for Biclustering Gene Expression Data

5.1 Introduction

Clustering techniques have been widely applied in gene expression analysis. However, clustering methods assume that related genes have the similar expression patterns across all conditions, which is not reasonable especially when the dataset contains many heterogeneous conditions. Recently, biclustering that simultaneously identifies groups of genes and groups of conditions over which the genes within a group exhibit similar expression patterns, has attained a lot of attention from researchers and practitioners. Biclustering is actually an old topic and was studied thirty years ago [Hartigan, 1975]. In 2000, Cheng and Church first introduced biclustering to gene expression analysis [Cheng and Church, 2000]. Many biclustering methods have been proposed for gene expression analysis such as δ-biclustering [Cheng and Church, 2000], coupled two-way clustering (CTWC) [Getz et al., 2000], statistical-algorithmic method for bicluster analysis (SAMBA) [Tanay et al., 2002], order-preserving biclustering [Ben-Dor et al., 2002, Liu et al., 2004], Plaid model [Lazzeroni and

Owen, 2002], spectral biclustering [Kluger et al., 2003], xMOTIF [Murali and Kasif, 2003], Gibbs-sampling-based biclustering [Sheng et al., 2003], flexible overlapped biclustering (FLOC) [Yang et al., 2003] , etc. See [Madeira and Oliveira, 2004] for an excellent survey.

In practice, gene expression data is often arranged as a matrix, where each gene corresponds to a row, each condition to a column and each element of the matrix represents the expression level of a gene under a specific condition. Thus, the goal of biclustering is to find one or more "homogeneous" submatrices that involve specific subsets of genes and conditions. [1] Basically, there are four major types of biclusters/submatrices: (i) biclusters with constant values, (ii) biclusters with constant values on rows or columns, (iii) biclusters with coherent values, and (iv) biclusters with coherent evolutions [Madeira and Oliveira, 2004]. So far, almost all proposed methods try to detect one or more types of biclusters by formulating it as an optimization problem. That is, a merit/objective function is employed to evaluate the quality of the seeked bicluster(s). The choice of the merit function is strongly related to the characteristics of the biclusters that each algorithm aims at finding. However, there is currently no proposed merit function that would allow an algorithm to find all types of biclusters. Hence, researchers have to design new merit functions in order to find new types of biclusters/patterns that are biologically interesting.

In this chapter, we propose a general merit for biclustering, which in principle can be used to detect any type of biclusters including the four aforementioned and any other possible computable patterns. The idea can also be easily applied to clustering. For biclustering, we actually try to find some submatrix of gene expression data with some interesting *regularity*. Previous methods have defined many kinds of regularities through different merit functions. To design a general framework for biclustering, we have to find a universal way to describe the regularity. Since *randomness* consists in the lack of regularity, it is natural for us to instead seek a universal way to measure the randomness of biclusters. In particular, we employ Kolmogorov complexity [Kol-

---

[1] Other formulations also exist. For example, the spectral biclustering method of Kluger *et al.* tries to find the checkerboard structure in the data through permutations on genes and conditions [Kluger et al., 2003].

mogorov, 1965, Kolmogorov, 1968, Kolmogorov, 1983a, Li and Vitányi, 1997, Solomonoff, 1964] to measure the randomness. The theory of Kolmogorov complexity is based on universal Turing machines. Informally, the Kolmogorov complexity of an object is the length of the shortest program for a universal Turing machine that correctly reproduces the object. Thus, Kolmogorov complexity is the ultimate measure of randomness and serves as a lower bound of the quantity of randomness that any real-world algorithm can possibly measure. In principle, one may also employ Shannon's information theory [Shannon, 1948] instead of the theory of Kolmogorov complexity to measure randomness. However, we prefer the theory of Kolmogorov complexity here because the probabilistic basis of Shannon's information theory requires us to know the distribution of gene expression data, which is usually unknown in practice. The drawback of Kolmogorov complexity is that it is uncomputable in general. In the method section, we will discuss how to approximate Kolmogorov complexity in the special case of gene expression data.

Based on the theory of Kolmogorov complexity, we formulate biclustering as an optimization problem in which the complexity density of a bicluster, i.e. the ratio of the Kolmogorov complexity of the bicluster to its size, is employed as the objective function to evaluate the quality of biclusters. To find optimal biclusters, we develop a Markov chain Monte Carlo (MCMC) algorithm based on a natural interpretation of conditional probability in algorithmic probability theory. Combined with the simulated annealing technique [Kirkpatrick et al., 1983], this method can effectively discover various patterns of interest as shown in some experiments on both simulated and real datasets.

The rest of the chapter is organized as follows. Section 5.2 gives a brief review of Kolmogorov complexity and algorithmic probability. In Section 5.3, we present the general framework for biclustering based on the theory of Kolmogorov complexity. We also discuss how to extend our method to the problems of clustering and checkerboard type biclustering. The approaches to approximate Kolmogorov complexity are also discussed in this section. Section 5.4 describe the experimental results on both simulated and real gene expression datasets. Section 5.5 concludes the chapter with some directions of further research.

5.2 Kolmogorov Complexity

Since the main idea behind this work is to detect regularity (lack of randomness), we will begin this section with a review of the algorithmic theory of randomness, which is the main root of the theory of Kolmogorov complexity. In what follows, we write string to mean a finite binary string. Other finite objects can be encoded as strings in a natural way. Given a long string x, the randomness of x naturally means that there is the absence of regularity in x [Kolmogorov, 1965]. Moreover, this string can be produced by a very simple program. Extending this argument, the non-randomness of x means that there is a program p producing x such that the length $|p|$ of p is much less than the length $|x|$ of x. With the formation rule of the circumference ratio π, for instance, we can write a simple program p that can generate any number of first digits of π although the sequence of digits of π seems very "random". This program p is actually the intrinsic descriptive complexity of π. In contrast, a really random string x cannot be generated by a program with length less than $|x|$ because the program has to explicitly print out the string without other choices. Consider a bicluster that exhibits some pattern/regularity. Based on our above discussion, this means that there is a "program" that can produce the expression levels of genes in a bicluster under specific conditions. This "program" could be a pathway or some transcriptional regulatory network. [2] In fact, the ultimate goal of biclustering is exactly to disclose the secrets of these "life programs".

The *Kolmogorov complexity* (also called *algorithmic entropy*) $K(x)$ of a string x is defined as the length of a shortest binary program p to compute x on an appropriate universal computer, such as a universal Turing machine. Thus, $K(x) = |p|$ (i.e. the length of p) denotes the minimum number of bits of information from which x can be computationally retrieved. If there are more than one shortest programs, then p is the first one in the standard enumeration. The *prefix Kolmogorov complexity* requires the programs of the universal computer to be prefix-free (no program is a proper prefix of another program). The difference between plain and prefix Kolmogorov complexity (and other variants) is only an

---

[2] Of course, the biological "programs" are not deterministic as computer programs and there are many variations and exceptions.

additive value of $\log |x|$. Prefix Kolmogorov complexity has a number of desirable mathematical characteristics that make it a more coherent theory. In this chapter, we use the prefix Kolmogorov complexity.

The conditional Kolmogorov complexity $K(x|y)$ of x relative to y is defined similarly as the length of a shortest program to compute x if y is furnished as an auxiliary input to the computation. The *algorithmic mutual information* is defined as $I(x : y) = K(y) - K(y|x, K(x))$ that is the information about x contained in y. We use the notation $K(x, y)$ for the length of a shortest binary program that prints out x and y and a description how to tell them apart. It is shown that there is a constant $c \geq 0$, independent of x, y, such that $K(x, y) = K(x) + K(y|x, K(x)) = K(y) + K(x|y, K(y))$ with the equalities holding up to c additive precision. Hence, up to an additive constant term, $I(x : y) = I(y : x)$.

It is easy to show that the functions $K(x)$ and $K(y|x)$ are upper semi-computable, but not computable. The algorithmic mutual information $I(x : y)$ is not even semi-computable. Since Kolmogorov complexity is not computable, we can only try to approximate it and relevant quantities such as the deficiency of randomness (to be discussed later) in applications.

Another important aspect of the theory of Kolmogorov complexity is the *algorithmic probability measure* [Solomonoff, 1964]. Given a universal computer $\mathcal{U}$, the algorithmic probability of a string x is

$$P_{\mathcal{U}}(x) = \sum_{p:\mathcal{U}(p)=x} 2^{-|p|} = Pr(\mathcal{U}(p) = x) \qquad (5.1)$$

which is the probability that a program randomly drawn as a sequence of fair coin flips $p_1, p_2, \ldots$ will print out the string x [Cover and Thomas, 1991]. This probability is universal in many senses and can be considered as the probability of observing such a string in nature. An important result is $P_{\mathcal{U}}(x) = 2^{-K(x)}$, where $K(x)$ is the *prefix* Kolmogorov complexity.

5.3 Methods

A bicluster with some pattern/regularity lacks randomness, which can be measured by the *deficiency of randomness* based on Kolmogorov complexity. First, we need note that the concept of randomness is relative because the quantity of the randomness of an object may change with respect to different backgrounds. Consider B that belongs to a set $\mathcal{B}$ (say, the set of all biclusters of size $n \times m$). The *deficiency of randomness* of B relative to $\mathcal{B}$ is defined as $\delta(B|\mathcal{B}) = \lceil \log |\mathcal{B}| \rceil - K(B|\mathcal{B})$, where $|\mathcal{B}|$ is the cardinality of $\mathcal{B}$. If $\delta(B|\mathcal{B})$ is large, then this means that there is a description of B with the help of $\mathcal{B}$ that is considerably shorter than the ordinal number of B in $\mathcal{B}$. In practice, the deficiency of randomness is often computed as $\delta(B) = |B| - K(B)$, i.e. $\mathcal{B}$ consists of all strings with the same size as B.

One may formulate the problem of biclustering as seeking a bicluster with the largest deficiency of randomness. However, such a formulation is not suitable because we would have to compare the deficiencies of randomness between two biclusters of different sizes when we look for optimal biclusters. Instead, we define the *complexity density*

$$h(B) = \frac{K(B)}{|B|} \tag{5.2}$$

as the merit function of biclusters because it removes the influence of the size of a bicluster. We can define some variants of the above merit function to search for some special biclusters. In order to favor large biclusters, for example, we may regularize the above function with the size of biclusters

$$h(B) = \frac{K(B)}{|B|} - \gamma |B| \tag{5.3}$$

where γ is the regularization factor that controls the extent to which the penalty term influences the size of a bicluster. For simplicity, we confine us to Equation (5.2) in the rest of chapter.

Formally, we formulate the problem of biclustering as follows. Given a gene expression dataset involving genes $g_1, g_2, \ldots, g_n$ and conditions

$c_1, c_2, \ldots, c_m$, we want to find a bicluster $B^* = (G, C)$ such that

$$B^* = \arg\min \frac{K(B)}{|B|} \qquad (5.4)$$

among all submatrices of the gene expression data, where G and C are two binary vectors whose elements g_i and c_j indicate if gene i and condition j are in the bicluster.

To solve this problem, we propose a Markov chain Monte Carlo (MCMC) algorithm that takes advantage of the Gibbs sampler and the algorithmic probability theory. The Gibbs sampler is a widely used MCMC scheme that follows the local dynamics of the target distribution [Geman and Geman, 1984]. Consider a random variable $\mathbf{x} = (x_1, \ldots, x_n)$. In the Gibbs sampler, we randomly or systematically choose a coordinate x_i and update it with a new sample x' drawn from the conditional distribution $\pi(\cdot|\mathbf{x}_{-i})$, where $\mathbf{x}_{-i} = (x_1, \ldots, x_{i-1}, x_{i+1}, \ldots, x_n)$. Although we do not know the target distribution $\pi(\cdot)$ for biclustering, we can easily interpret $\pi(\cdot|\mathbf{x}_{-i})$ in the algorithmic probability theory. Given the current bicluster B, the conditional probability $\pi(g_i|B_{-i})$ (or $\pi(c_j|B_{-j})$) tells us how probably the gene i (or condition j) belongs to the bicluster. A large $\pi(g_i|B_{-i})$ means that we can easily generate (or predict) the expression levels of gene i with the help of B_{-i}. In other words, the conditional complexity $K(g_i|B_{-i})$ is small. Thus, $\pi(g_i|B_{-i})$ can be estimated by the algorithmic probability $2^{-K(g_i|B_{-i})}$. This can be interpreted as a form of "Occam's razor": a gene whose expression can be easily predicted given the bicluster B_{-i} has a high probability.

Since $K(x, y) = K(x) + K(y|x, K(x))$, we have the approximation

$$2^{-K(g_i|B_{-i})} \approx 2^{-K(g_i|B_{-i}, K(B_{-i}))} = 2^{-(K(g_i, B_{-i}) - K(B_{-i}))}$$
$$= 2^{-K(g_i, B_{-i})} / 2^{-K(B_{-i})}$$

which can be thought of as the likelihood ratio by analogy. Note that we need normalize $2^{-K(g_i, B_{-i})}$ and $2^{-K(B_{-i})}$ because the pair (g_i, B_{-i}) has a larger size than the mere B_{-i}. A larger object potentially has greater complexity than a smaller object with the same structure. However, such excess complexity could mainly come from the additional part of the larger object. The *complexity density* in both objects may actually be the same. Suppose that the genes in the bicluster B_{-i} are independent

101

for simplicity, the algorithmic probability of complexity density will be $2^{-K(B_{-i})/|B_{-i}|}$, i.e. the geometric average of the algorithmic probabilities of genes. Thus, the normalized conditional probability $\pi(g_i|B_{-i})$ is

$$\pi(g_i|B_{-i}) = 2^{-(K(g_i,B_{-i})/|(g_i,B_{-i})|-K(B_{-i})/|B_{-i}|)} \qquad (5.5)$$

Note that $K(B_{-i})/|B_{-i}|$ in above equation is exactly our objective function (5.2).

With Equation (5.5), we have the following MCMC algorithm. Initially, we randomly assign some genes and conditions to $B^{(0)}$. Then we iteratively perform the following two steps. First, a gene/condition is randomly (in a uniform way) or systematically chosen and then added or removed from the bicluster to perturb the current bicluster $B^{(t)}$ to a new configuration B'. If it decreases the objective function $h(B)$, we will accept this perturbation and let $B^{(t+1)} = B'$. Otherwise, we accept it with probability $p = 2^{-\Delta h}$, where $\Delta h = h(B') - h(B^{(t)})$ represent the increase in $h(B)$. One may find that this procedure is the same as the Metropolis sampler. In fact, it is proven that the relaxed Gibbs sampler becomes the Metropolis sampler when the variables take only two possible values (e.g. in our biclustering formulation) [Liu, 1996].

To make the method more likely to find the global optimum, we combine the above algorithm with simulated annealing [Kirkpatrick et al., 1983]. In this procedure, a control parameter T, known as the "temperature", is introduced by an analogy to statistical mechanics. In the annealing procedure, we repeatedly perform the above algorithm with a monotonically decreasing temperature $T_1 > T_2 > \cdots > T_k > \cdots$. The complete algorithm is shown in Figure 5.1. Note that we replace the base 2 with e according to convention in the computation of accepting probability. The difference can be matched by adjusting T_k. In the algorithm, we also redefine $\Delta h = (h(B') - h(B^{(t)}))/l$, where l is the number of genes if a condition is selected to perturb the current bicluster or the number of conditions otherwise. Recall that the numbers of genes and conditions in biclusters are usually different. Thus, the changes in the objective function are also different when perturbing the bicluster with respect to genes or conditions. To be in the same favor of genes and conditions, we should normalize the change of objective function by the number of genes or conditions. It can be shown that the global optimum

> **ALGORITHM** A GENERAL FRAMEWORK FOR BICLUSTERING
>
> **Input:** A gene expression data matrix.
> **Output:** The boolean vectors G and C whose elements g_i and c_j indicate if the gene i and condition j are in the found bicluster B^*.
> **Method:**
>
> 1: Randomly assign some genes and conditions to the initial bicluster $B^{(0)}$.
> 2: Initialize temperature T_0.
> 3: **repeat**
> 4:   **for** a given number of iterations **do**
> 5:   Randomly or systematically select a gene i or a condition j.
> 6:   Generate a new bicluster B' by flipping g_i or c_j.
> 7:   Accept B', i.e. let $B^{(t+1)} = B'$, with probability $\min\{1, e^{-\Delta h/T_k}\}$, where $\Delta h = (h(B') - h(B^{(t)}))/l$ and l is the number of genes if a condition is selected to perturb the current bicluster or the number of conditions if a gene is selected. Otherwise, let $B^{(t+1)} = B^{(t)}$.
> 8:   **end for**
> 9:   Reduce temperature T_k to T_{k+1}.
> 10: **until** no change in the current bicluster.
> 11: Return the current bicluster as B^*.

Figure 5.1: A general framework for biclustering.

can be reached by this algorithm with probability 1 if the temperature T_k decreases sufficiently slowly [Geman and Geman, 1984].

This algorithm returns only one solution/bicluster each run. To obtain multiple biclusters, we may run the algorithm several times with different initial configurations. Besides, we may also employ the traditional masking technique to get multiple biclusters. That is, we mask the expression levels in the found bicluster with random numbers and then run the algorithm on the data again.

5.3.1 Clustering and Checkerboard Type Biclustering

We can easily modify the formulation of biclustering to solve the problem of conventional clustering. Consider the problem of clustering n genes in a gene expression data into k disjoint groups $G_i, i = 1, \ldots, k$. The objective is to minimize the total Kolmogorov complexity $\sum_{i=1}^{k} K(G_i)$ of all k clusters. We may still use the above algorithm to solve this problem by defining the configuration space as the set of vectors G of length n, whose ith element indicates which cluster gene i belongs to. Similarly, we can solve the problem of checkerboard type biclustering that simultaneously clusters genes into k_g groups and conditions into k_c groups. The objective function could be defined as $\sum_{i=1}^{k_g} \sum_{j=1}^{k_c} K(G_i, C_j)$. In these two formulations, we do not explicitly consider the sizes of the clusters in the objective function because the intrinsic constraints among the clusters (i.e. their sizes sum up to the size of the expression matrix) do the job.

5.3.2 Approximating Kolmogorov complexity

Recall that Kolmogorov complexity is not computable. Thus, we have to approximate it in practice. In this section, we discuss how to effectively approximate Kolmogorov complexity of gene expression. First, we review two widely used approximations, Shannon's entropy and Lempel-Ziv compressor. It is well-known that Shannon's entropy [Shannon, 1948] is a computable upper bound of Kolmogorov complexity. In [Kolmogorov, 1983a], Kolmogorov showed the following result. Let $x = x_1 x_2 \cdots x_n$, where each x_i takes s possible values and $p_i, i = 1, \ldots, s$ is the frequency of the occurrences of the ith value in x. Then for a large n,

$$K(x) \leq nH + s \log n \tag{5.6}$$

where $H = -\sum p_i \log p_i$ is Shannon's entropy. Note that this inequality holds only for a large n. Thus, we cannot detect the small biclusters with Shannon's entropy to approximate Kolmogorov complexity. Such a limitation makes entropy estimator impractical since most biclusters in real data are of small size. If the biclusters are sufficiently large, however, this approximation method works well and is not sensitive to noise as shown in the experiments.

The Lempel-Ziv (LZ) compressor family [Ziv and Lempel, 1977] is also widely used as an estimator of Kolmogorov complexity. When an LZ compressor encounters a phrase that has already been seen (saved in a codebook), it will generate a pointer to the match to replace the current phrase. Although LZ compressors work well in some applications, we observe that it is not suitable for our application due to over-compression. Figure 5.2 gives two examples of patterns that LZ compressors can achieve a high compression ratio. In the left subfigure, for example, the second row can be broken into two phrases "123" and "456" that can be matched with phrases in the first row. So, we can achieve a high compression ratio of about 0.5. However, the matching is not biologically meaningful. Besides, LZ algorithms are too slow to be used in an MCMC algorithm, which usually has a very large number of iterations.

Although entropy estimation and the LZ compression method look different, both of them are based on the frequencies of phrases. In entropy estimation, a phrase is the single expression level of a gene. On the other hand, a phrase is an arbitrary contiguous subsequence of expression levels in LZ compressors. Based on this observation, we may simplify the LZ-based estimation of Kolmogorov complexity and make it more biologically meaningful for gene expression analysis with the requirement that a phrase be a whole row or column of a bicluster. With this requirement, we can run the standard entropy estimation algorithm or LZ

1	2	3	4	5	6
4	5	6	1	2	3
...	...	...	...	...	...
...	...	...	...	...	...

1	2	3	4	5	6
7	8	1	2	3	4
5	6	7	8	1	2
3	4	5	6	7	8

Figure 5.2: Two examples of the over-compression of LZ compressors. In both cases, LZ algorithms can achieve a high compression ratio. However, the resultant patterns are not meaningful for gene expression analysis.

estimators for our application . In fact, we may go further. Note that every row/column in a *perfect* bicluster depicts the same biological process, which can essentially be represented by a *perfect* template row/column. On the other hand, we should not neglect the variances among the expression levels of genes when approximating the Kolmogorov complexity of biclusters because larger variances represent randomness while smaller variances imply better coherence of biclusters. Therefore, we may calculate the Kolmogorov complexity as

$$K(B) = K(t) + K(B|t) \qquad (5.7)$$

where t is a perfect template row/column. $K(t)$ is the complexity of the perfect template t, and $K(B|t)$ is the conditional complexity of the bicluster B given t. Because the value of $K(B|t)$ should somehow reflect the overall variances of rows/columns in B departing from the perfect template row/column t, we propose a simple heuristic algorithm to estimate it and a straightforward approach to construct the perfect template row/column t.

Assume that the bicluster under consideration is $B = (b_{ij})$, $1 \leq i \leq n$ and $1 \leq j \leq m$, and the template row is $t = (t_1, t_2, ..., t_m)$. Then, we set t_j to be the expression level that occurs the most frequently in the column j $\{b_{ij} : 1 \leq i \leq n\}$, i.e.

$$t_j = \arg \max_k \sum_{i=1}^{n} \delta(b_{ij}, k) \qquad (5.8)$$

where k is an expression level and $\delta(b_{ij}, k) = 1$ if $b_{ij} = k$, otherwise $\delta(b_{ij}, k) = 0$. Similarly, we can compute the template column. With this

106

1	1	1	1
1	1	1	1
1	1	1	1
1	1	1	1

1	1	1	1
2	2	2	2
3	3	3	3
4	4	4	4

1	3	4	2
3	5	6	4
6	8	9	7
5	7	8	6

7	13	19	2
19	23	39	6
4	6	8	2
3	7	8	1

Figure 5.3: Examples of bicluster with constant values, biclusters with constant rows, bicluster with coherent values (additive model), and bicluster with coherent evolutions (OPSM).

template row t, we calculate

$$K(B|t) = \sum_{1 \leq i \leq n} \sum_{1 \leq j \leq m} (1 - \delta(b_{ij}, t_j)). \qquad (5.9)$$

For simplicity, $K(t)$ can be estimated to be the size of t, i.e. $K(t) = m$, with the assumption that the expression levels of genes are independent. If $K(B|t)$ is defined as in Equation (5.9), it is easy to prove by contradiction that the template row/column t calculated as above is optimal in the sense that it results in the smallest value of Kolmogorov complexity $K(B)$ among all possible t. In Equation (5.9), we may also relax $\delta(b_{ij}, t_i)$ a little bit as

$$\tilde{\delta}(b_{ij}, t_i) = \begin{cases} 1 & \text{if } |b_{ij} - t_i| \leq \xi \\ 0 & \text{otherwise} \end{cases} \qquad (5.10)$$

where ξ is a positive number used as the threshold.

Equation (5.9) assumes a unit cost model. That is, if an expression level is different from the template, we use the same cost (i.e. 1) no matter how large the difference between the expression level and the template is. Such a simple cost model is only suitable when the background follows a uniform distribution. In practice, we will have to employ a more sophisticated cost model. More specifically, we use

$$K(B|t) = \sum_{1 \leq i \leq n} \sum_{1 \leq j \leq m} (\lceil \log |b_{ij} - t_j| \rceil + 1). \qquad (5.11)$$

That is, $\lceil \log |b_{ij} - t_j| \rceil$ bits are used to record the difference between b_{ij} and t_j, and the extra one bit is used to encode the sign of difference. For simplicity, we define $\log 0 = -1$ here.

In what follows, we show how to use the above method to approximate $K(B)$ for four important bicluster patterns: (i) biclusters with constant values, (ii) biclusters with constant rows/columns, (iii) biclusters with coherent values, and (iv) biclusters with coherent evolutions. Some examples of them are shown in Figure 5.3. Clearly, we can use the above discussed method to approximate the complexity of biclusters with constant rows/columns directly. We can also use the same procedure to approximate the complexity of biclusters with constant values since it is a special case of biclusters with constant rows/columns. In this case, the template t will be only a single expression level.

For biclusters with coherent values, let us consider the additive model used in Cheng and Church's δ-clustering method [Cheng and Church, 2000]. In the additive model, expression levels follow the rule as

$$b_{ij} = \mu + \alpha_i + \beta_j \qquad (5.12)$$

For this model, we need a simple preprocessing to use the above approximation method. Suppose the submatrix B has the size $n \times m$. To calculate the complexity, we generate another matrix S of size $n \times (m-1)$, where $s_{ij} = b_{i,j+1} - b_{ij}$, $i = 1, \ldots, n$ and $j = 1, \ldots, m - 1$. In S, each column has a constant value $\beta_{j+1} - \beta_j$. Then we compute the template t on the matrix S and $K(S|t)$ by Equations (5.8) and (5.9). Finally, we have

$$K(B) = K(B_{.1}) + K(S) = K(B_{.1}) + K(S|t) + K(t)$$
$$= K(S|t) + n + m - 1 \qquad (5.13)$$

where $B_{.1}$ is the first column of B, which is needed to recover B from S.

For biclusters with coherent evolutions, we consider a relaxed version of the order-preserving submatrix (OPSM) model in the method of Ben-Dor et al. [Ben-Dor et al., 2002]. A submatrix is order-preserving if there is a permutation of its columns under which the sequences of values in every row is strictly increasing. Note that this definition requires the order to be *globally* preserving between any two columns. In practice, the OPSM condition may be too strict. For simplicity, our relaxed OPSM model just requires that the order between two adjacent columns be preserving. In this case, we also generate a matrix S of size $n \times (m-1)$, where s_{ij} contains the order between $b_{i,j+1}$ and b_{ij}, $i = 1, \ldots, n$ and

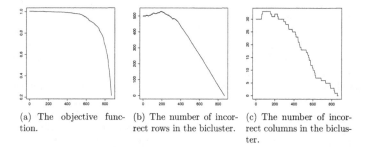

(a) The objective func-   (b) The number of incor-   (c) The number of incor-
tion.   rect rows in the bicluster. rect columns in the biclus-
                                                       ter.

Figure 5.4: The trace of the algorithm on a uniform bicluster. The x-axis is the number of successful moves in the MCMC algorithm.

$j = 1, \ldots, m - 1$. Similar to the case of additive model, we can calculate $K(B)$ by Equation (5.13).

5.4   Experiments

We have implemented our method and run it on simulated as well as real data. To test if our method can detect different types of biclusters, we generate a large number of simulated datasets, all of them are 1000×50 matrices and each is embedded with a different type of biclusters. Here, we report the results on four typical kinds of biclusters: biclusters with constant values, biclusters with constant rows, biclusters of additive model, and biclusters of relaxed OPSM. All of these biclusters have size 100×20. The dataset generating procedure is as follows. First, the data matrix is filled with random values uniformly distributed in the range $(20, 16000)$. Then a target bicluster perturbed by random noise is embedded into the data matrix. Finally, we permute the rows and columns of the data matrix so that the bicluster does not appear as a contiguous submatrix.

Since gene expression data have real values, we need discretize it to approximate Kolmogorov complexity. Because gene expression data is usually highly noised, the discretization may also help us to reduce the influence of noise in practice. In the experiments, we discretize the expression levels of genes with a combination of equal frequency and equal width discretization methods. Namely, the range of expression levels is first divided into k intervals that each contains a (roughly) same number of continuous values. If the length of an interval is too long, the interval will be divided into several smaller ones with equal width. Finally, all expression levels in the ith interval are assigned the value i. The reported results are based on $k = 128$. Other choices of k such as 64 and 256 give similar results. Because the background follows a uniform distribution, we use the relaxed unit cost model as given in (5.10) with $\xi = 1$ to approximate Kolmogorov complexity in this experiment. For the annealing procedure, it is known that the global optimum can be reached by the simulated annealing algorithm with probability 1 if the temperature decreases sufficiently slowly (i.e. at the order of $O(\log(L_k)^{-1})$, where $L_k = N_1 + \cdots + N_k$ and N_k is the number of iterations at the temperature T_k) [Geman and Geman, 1984]. In practice, however, no one can afford such a slow annealing schedule. As most people did, we use the exponential cooling scheme, which was proposed by Kirkpatrick et $al.$ [Kirkpatrick et al., 1983]. That is, $T_k = \lambda T_{k-1}$ where λ is a constant close to, but smaller than 1. Although such an annealing schedule does not theoretically guarantee that the global optimum can be reached, we find that it actually works very well in our biclustering application as shown later. For all simulated data, we set the initial temperature to $1E - 5$ and $\lambda = 0.95$ in the experiments. We also randomly choose about a half of the genes and conditions as the initial configuration of the MCMC algorithm. It is interesting that our method can detect all four types of biclusters without any error with the above setting. That is, all genes and conditions in the target biclusters are successfully detected and no other genes/conditions are included in the solutions. In Figure 5.4, we give the trace of objective function on a uniform bicluster. We also depict the difference between the true bicluster and the found bicluster during the iterations. The trace on other types of biclusters has the similar pattern. As shown in the figure, our method can converge to true biclusters of all types without any error. Besides, the algorithm converges fast and returns the solution in less than 1000 successful moves (The total number of iterations is set to 105000). Because the operations in each iteration are very simple, the algorithm usually terminate in less than one minute on these datasets.

110

Table 5.1: The results of the δ-biclustering and SAMBA algorithm on the biclusters with constant values, constant rows, and additive model.

	δ-biclustering						SAMBA					
	Constant values		Constant rows		Additive model		Constant values		Constant rows		Additive model	
	δ	size	δ	size	δ	size	score	size	score	size	score	size
	1E+5	5 × 10	1E+6	5 × 10	1E+8	5 × 10	54.76	18 × 4	55.51	18 × 4	54.12	20 × 4
	5E+5	11 × 10	5E+6	7 × 10	5E+8	11 × 10	53.68	21 × 4	53.50	21 × 4	39.59	14 × 5
	1E+6	20 × 15	1E+7	19 × 12	1E+9	18 × 14	40.71	14 × 5	41.01	14 × 5	63.83	14 × 6
	5E+6	68 × 20	5E+7	44 × 20	1.5E+9	28 × 22	46.28	19 × 4	45.87	19 × 4	60.07	21 × 4
	1E+7	69 × 20	1E+8	45 × 20	2E+9	47 × 29	57.92	22 × 4	56.63	21 × 4	39.96	10 × 4
	5E+7	73 × 22	5E+8	52 × 21	2.5E+9	75 × 43	39.41	11 × 4	39.49	10 × 4	59.51	20 × 4
	1E+8	76 × 28	1E+9	59 × 24	3E+9	213 × 50	59.52	20 × 4	59.42	20 × 4	62.85	23 × 4
	5E+8	1000 × 50	5E+9	1000 × 50	3.5E+9	1000 × 50	61.78	21 × 4	61.97	21 × 4	55.61	20 × 4
	N/A	N/A	N/A	N/A	N/A	N/A	55.65	18 × 4	55.52	18 × 4	47.81	13 × 5
	N/A	N/A	N/A	N/A	N/A	N/A	46.49	13 × 5	46.99	13 × 5	66.79	24 × 4
	N/A	N/A	N/A	N/A	N/A	N/A	61.98	19 × 4	61.98	19 × 4	61.83	21 × 4
	N/A	N/A	N/A	N/A	N/A	N/A	60.68	21 × 4	61.45	21 × 4	63.69	21 × 4
	N/A	N/A	N/A	N/A	N/A	N/A	63.44	20 × 4	63.59	20 × 4	52.35	16 × 4
	N/A	N/A	N/A	N/A	N/A	N/A	53.13	15 × 4	52.89	15 × 4	210.99	38 × 6
	N/A	N/A	N/A	N/A	N/A	N/A	45.42	14 × 4	45.46	14 × 4	210.29	26 × 9
	N/A	N/A	N/A	N/A	N/A	N/A	49.74	17 × 4	49.74	17 × 4	N/A	N/A

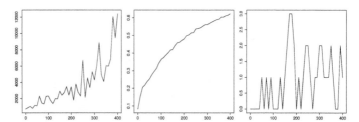

Figure 5.5: The number of successful moves (left), objective function value (middle) and incorrect rows (right) when biclusters are noisy. The incorrect columns are always zero in all cases and thus are not shown here. The x-axis is the standard deviation of noise.

As a comparison, we also run the δ-biclustering algorithm of Cheng and Church [Cheng and Church, 2000] [3] and the SAMBA algorithm of Tanay *et al.* [Tanay et al., 2002] [4] on these datasets. For δ-biclustering, we try many different threshold δ and list the results in Table 5.1. On the bicluster with constant values, the best result of δ-biclustering is a bicluster of size 73×22, of which 65 rows and 20 columns are correct. The reported H score [Cheng and Church, 2000] is 40543429, which is far from that (78203) of the true bicluster. As shown in the experiments, the threshold δ plays a crucial role. When δ is large, say 5E+8, the whole data matrix is returned. On the other hand, a very small bicluster of size 5×10 is returned when $\delta = 1E + 5$. Note that 1E+5 is actually close to the H score of the true bicluster. The results on the bicluster of constant rows and additive model are similar. For SAMBA, we use the recommended setting valsp_3ap. The reported biclusters are listed in Table 5.1. We observe that all reported biclusters are small with respect to the true biclusters. The reason may be that the polynomial algorithm of SAMBA requires that the degree of every gene vertex is bounded.

As mentioned before, our method with entropy estimator is very robust to the noise. To test it, we perform a series of experiments on biclusters with constant values. In the experiments, the sizes of biclusters and data matrices are still 100×20 and 1000×50, respectively. We

---

[3]http://cheng.ececs.uc.edu/biclustering/Biclustering.java

[4]http://www.cs.tau.ac.il/~rshamir/expander/expander.html

generate biclusters with constant expression level 5000. The expression levels are perturbed by random numbers drawn from the Gaussian distribution $N(0, \delta^2)$, where the standard deviation δ increases from 10 to 400 by 10 each step. Note that the standard deviation 400 is considerably large compared to the expression level 5000. In all experiments, the discretization parameter k is set to 128 and T_0 is set to $5E - 5$. As shown in Figure 5.5, our method is very robust to noise. In all cases, the method could correctly detect all conditions in biclusters without any error. With respect to the genes, the maximum number of errors is only 3, which is very small relative to the total number (i.e. 100) of genes in the target biclusters. On the other hand, we have observed that the number of successful moves and the objective function value of the found bicluster increase with the degree of noise in general.

Besides simulated datasets, we also run our algorithm on a real yeast cell cycle gene expression data produced by Spellman *et al.* [Spellman et al., 1998]. Spellman *et al.* measured the relative expression levels of 6218 *Saccharomyces cerevisiae* putative gene transcripts (ORFs) as a function of time in cell cultures that had been synchronized in three independent ways, α factor-based synchronization, size-based synchronization, and cdc-15 based synchronization, which results in a dataset containing 77 conditions (time points). Since the data involve many different conditions, it is a good benchmark data to test our biclustering method. Spellman *et al.* identified 799 genes that are cell cycle regulated. We use the expression levels of these 799 genes in the experiments.

With Equation (5.11) to approximate conditional Kolmogorov complexity, our biclustering method finds many interesting biclusters on the data by setting the initial temperature to $1E - 7$ and $\lambda = 0.95$. Some of them are shown in Figure 5.6. These biclusters follows either the additive model or relaxed OPSM model. Clearly, genes in the same biclusters show the similar expression patterns. Our method also finds many biclusters following uniform or constant rows model, we do not show them here because most uniform and constant-row biclusters are just associated primarily with groups of genes that do not respond in the tested experiments and are not with some biologically significant phenomenon. By identifying and masking these "background" uniform and constant-row biclusters, however, we may detect biologically interesting bicluster more easily. In fact, the biclusters of additive model shown in Figure 5.6 are found after we randomly mask 30 detected uniform biclusters.

To determine the statistical significance for functional category enrichment, we use the hypergeometric distribution to model the probability of observing at least k ORFs from a cluster size n by chance in a category containing f ORFs from a total genome size of g ORFs ($g = 6613$ in current MIPS database) [Tavazoie et al., 1999]. More specifically, the P-value is given as

$$P = 1 - \sum_{i=0}^{k} \frac{\binom{f}{i}\binom{g-f}{n-i}}{\binom{g}{n}} \tag{5.14}$$

which can be used to measure if a bicluster is enriched with genes from a particular functional category to a greater extent than that would be expected by chance. If the majority of ORFs in a bicluster belong to one functional category, it is unlikely to happen by chance and the P-value will be close to 0. We test the third level MIPS functional categories [Mewes et al., 1997, Ruepp et al., 2004] for each bicluster. By adopting Bonferroni's correction for multiple independent hypotheses, only P-values less than 2×10^{-4} are reported because otherwise the total expectation within the bicluster would be higher than 0.05. As shown in Table 5.2, most biclusters are enriched by genes from some particular functional categories. Note that many genes belong to multiple functional categories. So, the summation of the number of genes within each functional category could be greater than the total number of genes in the bicluster.

We then search for upstream DNA sequence motifs that are common to members of the above biclusters in order to identify known or novel *cis*-regulatory elements that may contribute to the co-regulation of genes in a bicluster [Tavazoie et al., 1999]. We use the program AlignACE 3.0 [Roth et al., 1998] to search for DNA sequence motifs in the 600bp regions upstream of the ORFs in biclusters 7 and 8. For AlignACE, we set both the expected number of conserved bases and the expected number of sites to 10. Two motifs found in the biclusters are graphically represented in Figure 5.7 by the program WebLogo [Crooks et al., 2004]. Note that both motifs have the considerably large MAP (maximum *a priori* log likelihood) scores (62.663 and 43.0691, respectively), which means that the motif is highly over-represented relative to the

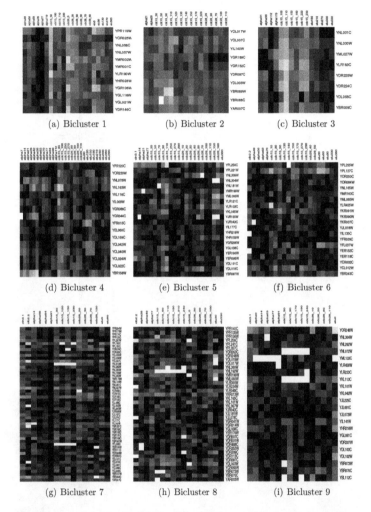

(a) Bicluster 1   (b) Bicluster 2   (c) Bicluster 3

(d) Bicluster 4   (e) Bicluster 5   (f) Bicluster 6

(g) Bicluster 7   (h) Bicluster 8   (i) Bicluster 9

Figure 5.6: Some biclusters found in the yeast cell cycle gene expression data of Spellman *et al.* [Spellman et al., 1998]. The biclusters 1-6 are from the additive model and the biclusters 7-9 are from the relaxed OPSM model.

115

Table 5.2: Enrichment of biclusters within functional categories.

Bicluster	Number of ORFs	Number of conditions	MIPS functional category (total ORFs)	ORFs within functional category	P-value $-log_{10}$
1	12	21	mitotic cell cycle and cell cycle control (442)	6	5
3	8	15	DNA restriction or modification (196)	5	6
			mRNA synthesis (572)	7	7
			DNA binding (159)	6	9
5	22	23	DNA synthesis and replication (179)	8	8
			DNA recombination and DNA repair (246)	8	7
7	52	16	DNA synthesis and replication (179)	8	16
			DNA recombination and DNA repair (246)	18	16
			mitotic cell cycle and cell cycle control (442)	17	12
			DNA damage response (77)	15	7
8	40	17	mitotic cell cycle and cell cycle control (442)	11	7
			chemoperception and response (265)	8	5
			fungal and other eukaryotic cell type differentiation (449)	11	5
9	22	16	DNA synthesis and replication (179)	8	8
			DNA damage response (77)	4	4

(a) Motif found in bicluster 7 with MAP score 62.663.

(b) Motif found in bicluster 8 with MAP score 43.0691

Figure 5.7: Motifs found in biclusters 7 and 8. The MAP score is a statistic given by AlignACE to determine the significance of an alignment. A large MAP score (say 10) indicates the high significance of motif.

expectation for the random occurrence of such a motif in the sequence under consideration. In fact, both motifs are also biologically significant because they match experimentally verified transcription factor binding sites well. For example, the motif in Figure 5.7(a) matches the MCB binding site (ACGCGT) and the motif in Figure 5.7(b) matches the SCB binding site (CACGAAA) well. Besides co-regulation, genes may be co-expressed due to other reasons as we found in bicluster 3. This bicluster contains eight genes that are YBR009C, YDL055C, YDR224C, YDR225W, YLR183C, YML027W, YNL030W, and YNL031C. Interestingly, the products of five genes (YBR009C, YDR224C, YDR225W, YNL030W, and YNL031C) participate in the nucleosomal protein complex (MIPS website, http://mips.gsf.de/), which may be the main reason of their co-expression.

5.5 Conclusion

In this chapter, we have proposed a novel biclustering method based on Kolmogorov complexity. Instead of pairwise similarity between gene expressions, our method considers the overall coherence of biclusters. More precisely, we detect biclusters by minimizing the complexity density by using an MCMC algorithm. Our method is general in that it can discover any type of computable patterns in gene expression data. Our experimental results on simulated and real data show that the approach is very versatile and promising.

Chapter 6

Systematic Discovery of Functional Modules and Context-Specific Functional Annotation of Human Genome

6.1 Introduction

Systematic functional characterization of genes identified in the genome sequencing projects is urgently needed in the post-genomic era. The rapid increase in large-scale gene expression data provides us unique opportunities to meet this need. A commonly used approach is to cluster genes with similar expression patterns [Beer and Tavazoie, 2004, Gasch and Eisen, 2002, Tamayo et al., 1999], and to predict functions of unknown genes based on their expression-similarity to known genes [Gasch and Eisen, 2002, Niehrs and Pollet, 1999]. However, there are two problems with such clustering approaches: (Problem 1) Genes with similar expression profiles may not have the same function: For example, an experimental condition may perturb multiple biological pathways simultaneously, such that genes from these different functional pathways may show similar and indistinguishable expression patterns; and moreover, experimental noise and outliers may lead to biased and erroneously high estimates of expression similarity. (Problem 2) Genes with similar functions may not have similar expression profiles: For example, measurements of expression similarity, e.g. Pearson's correlation or Euclidean distance, may not capture the relationship between two expression profiles due to time-shifts [Qian et al., 2001]; and genes may be regulated at levels other than transcription. Recently, two novel expression relationships, "transitive expression similarity" [Zhou et al., 2002] and "second-order expression similarity" [Zhou et al., 2005] have been proposed and validated. These methods can be used to link functionally related genes without similar expression profiles. No doubt that many other types of expression relationships exist among functionally related genes – some may even be beyond our current knowledge. How to identify such unknown expression relationships systematically is one of the major aims of this chapter. In addition, we will address another important problem (Problem 3) in functional annotation, which has so far received little attention: how to annotate gene functions in a context-specific manner? An increasing number of examples indicate that in higher organisms, functional plasticity may be the rule rather than the exception [Jeffery, 2003a, Jeffery, 2003b]. A gene may acquire different functions under different endogenous or exogenous conditions. However, current functional prediction approaches [Wu et al., 2002, Zhu et al., 2005] and genome databases (such as SGD [Wang et al., 2005], Wormbase [Drysdale et al., 2005], FlyBase [Aggarwal et al., 2006]) all annotate gene functions without specifying the necessary context. Recently, Lussier et al. for the first time systematically addressed this problem by proposing a system, PhenoGO, which extracts phenotypic contextual information from published literatures for existing GeneOntology functional annotations [Lussier et al., 2006].

In this chapter, we aim to overcome the above three issues by using information in multiple microarray data sets. We model each microarray data set with a graph, where a vertex represents a gene, and if two genes show high correlation in their expression profiles, we connect them with an edge. A series of microarray data sets can be modeled as a series of co-expression networks, in which we search for frequently occurring network patterns. Such a network pattern consists of gene sets that function as a unit under various conditions, and thus likely represent a functional module. Based on the recurrent network patterns, we perform functional annotation. This approach can address the afore-mentioned problems: (1) To separate true functional links from spurious co-expression links. We suggest that a co-expression link recurrent in multiple microarray data sets is more likely to represent a true func-tional link. (2) To identify functionally related genes without direct co-expression. When we combine multiple expression networks, subtle signals may emerge that cannot be identified in any of the individual networks. Such signals include recurrent paths that may extend beyond simple co-expression clusters yet represent functional modules. If we only consider a single co-expression network, it is difficult to stratify function-ally important paths from their complex network environment. However, if a path frequently occurs across multiple co-expression networks, it is easily differentiated from the background. (3) To conditionally annotate gene functions. Because a gene can not exert its function by itself but instead does so by interaction with other genes, its functional switch is likely to be caused by or result from the alteration of its interaction part-ners. Put into a network perspective, a gene's function may be different if placed in different subnetworks. As different external or endogenous conditions result in different topologies of the co-expression network, we can relate a gene's function to the experimental conditions via its network environment, thus leading to the context-specific functional annotation.

In this study, we integrate 65 human microarray data sets, com-prising 1105 experiments and over 11 million expression measurements. We develop a data mining procedure based on frequent itemset min-ing and biclustering to extensively discover network patterns that re-cur in at least 5 datasets. This resulted in 143,401 potential func-tional modules. Subsequently, we design a network topology statistic based on graph random walk that effectively captures characteristics of a gene's local functional environment. Functional annotations based on this statistic are then assessed using the random forest method with six other attributes of the network modules. We assign 1126 functions to 895 genes, 779 known and 116 unknown, with a validation accuracy of 70%. Note that predictions on known genes were used only for val-idation in previous studies. Our predictions on known genes, on the other hand, additionally provide the context information of genes' func-tion. Among our assignment, 20% genes are assigned with multiple func-tions based on different network environments. The functional predic-

tions together with the necessary context information are available at
http://zhoulab.usc.edu/ContextAnnotation.

6.2 Materials and Methods

6.2.1 Microarray Data

We collected 65 human microarray datasets including 52 Affymetrix
(U133 and U95 platforms) and 13 cDNA datasets (details see Supplementary website) from the NCBI GEO [Edgar et al., 2002] and SMD [Gollub
et al., 2003] databases (version December 2005). The selection criteria
are that each dataset contains at least eight experiments and that the
percentage of statistically significant co-expressed gene pairs (see the section of graph construction for details) is not higher than 3%. The first
criterion ensures that the dataset contains enough profiles so that the
constructed co-expression graph is reliable while the second criterion filters out the datasets, in which too broad perturbations result in a large
number of spurious co-expression estimates. The collected datasets are
preprocessed as follows. The datasets generated from Affymetrix chips
are log-transformed (base e) to place them on the same scale as the
cDNA datasets. Note that the original values less than 10 in Affymetrix
datasets are set to 10. For each dataset, genes with low expression variation (lowest 10% in terms of the ratio of standard deviation to mean
for Affymetrix data and of standard deviation for cDNA data.) are discarded. Finally, genes with more than 30% missing values and arrays
with more than 20% missing values are also discarded.

6.2.2 Gene Ontology Function Categories

The Gene Ontology file on Biological Processes was downloaded from
GO consortium (Feb. 2005). The associated annotation information for
human genes was from the NCBI Gene Database (Aug. 2005). With the
method proposed in [Zhou et al., 2002], we selected process categories

from GO that contain more than 175 genes but each of their children contains less than 175 genes. The 40 GO categories obtained are called the informative functional categories and will be used in functional annotation later. Note that genes with annotations but not belonging to any of the informative categories were discarded. After the data preprocessing, the 65 datasets comprise in total 8297 genes, of which 5629 have at least one known function and 2668 do not have any known functions.

6.2.3 Graph Construction

Each microarray dataset is modeled as a relation graph where each node represents one gene and two genes are connected if their expression correlation is significant. Here the expression correlation, denoted as r, is taken as the minimum of the absolute value of leave-one-out Pearson correlation coefficients, which is robust against single experiment outliers and sensitive to overall similarities in expression patterns [Zhou et al., 2002]. We then use the statistic $t = \sqrt{(n-2)r^2/(1-r^2)}$ to determine if the expression correlation is significant. More precisely, the quantity t is modeled as a t-distribution with $n - 2$ degrees of freedom, where n is the number of measurements used in the computation of the correlation. In our study, an expression correlation significant at $P \leq 0.01$ level is included as an edge in the relation graph.

6.2.4 Mining Recurrent Network Modules

Given 65 graphs, each of which contains (at most) 8297 nodes, we attempt to identify connected network patterns that comprise at least 4 nodes and that occur in at least 5 graphs. This is computationally very difficult due to the exponential number of potential patterns. In our approach, we first search for frequent edge sets that are not necessarily connected and then extract connected components from them. Conceptually, we represent the 65 graphs as a matrix where each row represents an edge (i.e. a gene pair), each column represents a graph, and each entry (0 or 1) indicates whether the edge appears in that graph. Clearly, our problem of discovering frequent edge sets can be formulated as a typical biclustering problem that searches for submatrices with high density of 1s, which is a well-known NP-hard problem.

We developed a biclustering algorithm based on simulated annealing to discover frequent edge sets. More precisely, we employ simulated annealing to maximize the objective function $\frac{c'}{mn+\lambda c}$, where c is the number of 1s in the input matrix, c', m and n are the numbers of 1s, rows and columns of the bicluster, respectively, and λ is a regularization factor. Clearly, such an objective function is in favor of biclusters with a high density of 1 and with large size. Note that the density is maximized to 1 when $c' = mn$, while the size of bicluster is maximized when $c' = c$ (i.e. the pattern is as large as the input matrix). The regularization parameter λ controls the compromise between the density and the size. However, there is no theoretic result on selecting optimal λ. In the study, we tried many heuristic choices of λ and the reported results are based on $\lambda = 0.2/\max(1, \log_{10} n_1)$, where n is the number of edges of the initial configuration (i.e. seed).

Although this method performs well in our experiments, the search space has to be restricted in order to discover hundreds of thousands patterns in reasonable time because of the huge size of matrix (more than one million rows and 65 columns). As an attempt to solve this problem, we employ the frequent itemset mining (FIM) technique [Grahne and Zhu, 2003] to restrict the search space and also provide seeds for our biclustering algorithm. In what follows, we briefly describe FIM and related concepts. Let $I = \{i_1, i_2, \ldots, i_n\}$ be a set of items and database D be a set of m transactions, where each transaction T is a set of items such that $T \subseteq D$. Let X be a set of items. A transaction T is said to contain X if and only if $X \subseteq T$. X is frequent if at least S transactions in the database contain X. In our case, we regard an edge as a transaction and its occurrence in a particular graph as an item. For our purpose, we include only edges occurring in at least 5 graphs in the transaction dataset. The output of FIM algorithms is the set of all possible item sets which occur in at least S transactions. Note that the submatrices of frequent itemsets and their supporting transactions are actually biclusters full of 1s. These clusters with perfect density can then serve as seeds for our biclustering algorithm to search for larger biclusters that permit holes (i.e.

0s). We ended up with about 1.8 millions frequent itemsets which contain at least 4 edges and occur in at least 5 graphs. These FIM patterns, however, should not be used as seeds directly because they are highly overlapping, which is due to the nature of frequent itemset's definition. This is well-known in data mining community. In order to improve these seeds and also reduce unnecessary computation in final biclustering, we first remove FIM patterns whose supporting transactions/edges are the subset of those of other patterns. Second, we also merge two patterns if their union has a density larger than 0.8. This procedure is repeated until no additional merge can happen.

After the post-processing, we finally have about half million merged FIM patterns to feed our biclustering algorithm. Given a FIM pattern with v member genes, we will use all possible $v(v-1)/2$ edges among these v genes and all datasets as the input matrix for our biclustering algorithm. The FIM pattern is also used as the initial configuration of simulated annealing. From the output biclusters of our algorithm, we extract connected components as the final output patterns.

6.2.5 Network Topology Score for Each Function Category

Given a network pattern, the most popular gene function prediction method involves the use of the hyper-geometric distribution to model the probability of genes function based on neighborhood. This method however ignores the network topology, which is probably the most important information in the network patterns. To avoid this problem, we developed a new method based on graph random walk to fully explore the topology of network patterns. Our method is still based on the principle of "guilt by association". In terms of network topology, the association between genes is measured by how close they are (i.e. the length of path between them) and how tightly connected they are (i.e. how many paths between them). Statistically, this translates to how likely it is to reach one gene starting from another gene in a random walk. This probability can be approximately calculated by matrix multiplication.

Given a network pattern consisting of v genes, Let P be a stochastic matrix of size $v \times v$, of which the element P_{ij} is $1/n_i$ if genes i and j are connected, or 0 otherwise, where n_i is the number of neighbors of gene i. If we regard genes as states and P_{ij} as the probability of transformation from genes/states i to j, then the random walk on the graph can be thought as a Markov process. Therefore, it is easy to see that the element of P^k is the probability that gene i reaches gene j in k steps of the walk. The intuition behind our method is that genes with similar function are more likely to be well-connected (i.e. gene i can reach gene j with high probability in a random walk). Simply put, the probability P_{ij}^k would be large if genes i and j share the same function. Let O be the Gene Ontology binary matrix of which element O_{ij} is 1 if gene i belongs to category j and 0 otherwise. Thus, the matrix $M = P^k O$ gives the scores of genes relating to functional categories. The higher the score, the more likely a gene has that function. In practice, we choose $k = 3$ because we would like to confine our prediction to a local area of network patterns. With the score matrix M, the function of each gene is estimated by finding the functional category with the maximum score in the corresponding row of M.

6.2.6 Assessment of Function-Assignments with Random Forest

In an attempt to improve our method, we wanted to include other attributes of network patterns in the final prediction beside the network topology score. In particular, we include recurrence, density, size, average node degree, percentage of unknown genes, and functional enrichment of network modules. To take those factors into account, we use random forest [Breiman, 2001] to determine whether the function assignments based on the network topology score is robust. Note that the purpose of the random forest here is to determine whether to accept or reject the functional assignment made based on the network topology score. The random forest was trained using the assignments of known genes. The trained model was then applied to assignments of unknown genes. Finally, we keep only the function assignments that the random forest classified as "accept".

6.3 Results

6.3.1 Systematic Identification of Functional Modules in Human Genome

We constructed 65 co-expression networks from 65 microarray datasets. In total, the graphs contain 8297 genes. Using the graph-based mining approach as described in Methods, we obtained 1,823,518 network patterns which occur in at least 5 graphs. We further designed a biclustering approach (see Methods) to merge patterns similar in both their network topology and their dataset recurrence. This drove down the number of network patterns to 143,400, which covers 2769 known and 1054 unknown genes and varies in size from 4 to 180. The whole pipeline took 254 CPU hours (2GHz AMD Opteron Processor 270). In general, the size of a module is inversely proportional to its recurrence. Among those modules, 45% of modules each contain more than 90% known genes, which allow us to assess the functional homogeneity of the module. We define a module to be functionally homogenous if the hyper-geometric P-value, after Bonferroni correction, is less than 0.01. Among the identified network modules, 77.0% of the patterns are functionally homogenous.

Figure 6.1(a) shows the histogram of network recurrence across the 65 datasets, which approximates an exponential distribution. We define a module to be active in a dataset if 80% of its edges appear in that dataset. The recurrence of those modules ranges from 5 to 20. The most frequently occurring module (with recurrence 20) contains five genes MT1E, MT1F, MT1H, MT1L, MT1X (the network topology is shown in Figure 6.2(a)), all of which are metallothionein and only MT1X has known annotations in the GO database as "nitric oxide mediated signal transduction" and "response to metal ion". Multiple experimental studies have revealed concurrent activities of the MT genes in intracellular defense against reactive oxygen and nitrogen species. For example, substances causing oxidative stress and agents involved in inflammatory processes induce the synthesis of metallothionein [Chun et al., 2004, Chung et al., 2006, Izmailova et al., 2003]. This evidence is consistent with their observed tight expression regulation in our study. Another frequent occurring

127

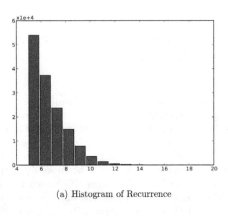

(a) Histogram of Recurrence

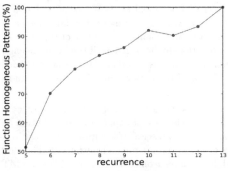

(b) Function Homogeneity vs Recurrence

Figure 6.1: (a) Distribution of recurrence of identified modules approximates an exponential distribution. (b) For all modules containing 7 genes, the percentage of function-homogeneous modules increases wih recurrence. Modules of different sizes show similar trend (see Supplementary website).

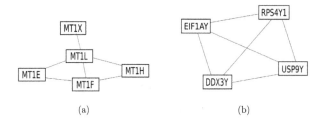

(a)	(b)

Figure 6.2: Two network modules with high recurrences.

module, comprising four genes RPS4Y1, USP9Y, DDX3Y, EIF1AY as a clique (the network topology is shown in Figure 6.2(b)), occurs in 18 datasets. Interestingly, USP9Y, DDX3Y, EIF1AY are all located in the chromosomal region Yq11 and considered to be involved in spermatogenesis [Vogt, 2005]. In addition, RPS4Y1 shares high sequence similarity with RPS4Y2, which also resides in Yq11 and is linked to spermatogenesis. These examples demonstrate that recurrent network modules are highly likely to be involved in a specific biological process.

In general, the higher the recurrence, the more likely the modules are to be functionally homogenous (Figure 6.1(b)). This lays the foundation for using multiple microarray datasets to enhance the functional inferences. In fact, when the recurrence is high, even loosely connected network patterns, or paths, can represent functional modules. We define the connectivity of a graph g to be $\frac{2m}{n(n-1)}$, where m is the number of edges and n the number of vertices in g. Figure 6.3(a) shows an example. All seven genes are involved in "immune response", though extremely loosely connected; they are identified through their occurrence in 6 graphs. Most current algorithms identify network modules by looking for densely connected subnetworks in a single network [Bader and Hogue, 2003, Shannon et al., 2003, Spirin and Mirny, 2003]. Here, by considering pattern recurrence across many networks, we are able to identify network modules of most topologies. In fact, 24% of the identified modules have connectivity less than 0.5. Figure 6.3(b) shows the network connectivity distribution of the modules.

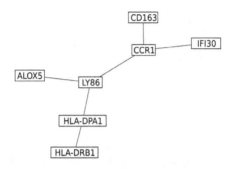

(a) A loosely connected network module

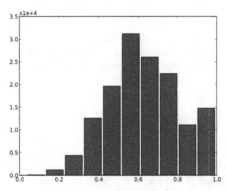

(b) Histogram of Connectivity

Figure 6.3: (a) A loosely connected network module with connectivity 0.28 enriched with the function "immune response" (P-value$< 10^{-7}$). (b) Distribution of network connectivity among identified modules.

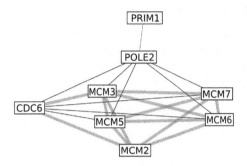

Figure 6.4: A network pattern enriched with protein interaction. Edges representing protein interaction are colored in red.

6.3.2 Enrichment of Protein-Protein Interaction in Network Modules

To explore the types of interaction relations among the network members beyond co-expression, we resort to the only available large-scale interaction information, protein interaction data. We retrieved the human protein interaction information from EBI(European Bioinformatics Institute)/IntAct [Hermjakob et al., 2004] (version 2006-10-13), and test for each of the 143,400 modules whether protein interaction was over-represented among its member genes compared to all human genes based on the hyper-geometric test. 60556 (22.44%) network modules were found to be enriched in protein interaction at P-value 0.001 level. This shows that genes in our network modules are more likely to encode interacting proteins. Interestingly, many of the protein-interaction enriched network modules fall into functional categories, such as protein biosynthesis, DNA metabolism, etc, and many interacting protein pairs are not necessarily co-expressed. Figure 6.4 shows an example of such network modules with two edge colors indicating co-expression and protein interaction, respectively.

6.3.3 Function Prediction

Based on the 143,400 recurrent network patterns, we assigned to each gene the function with maximum network topology score in each network pattern and made functional predictions for 779 known and 116 unknown genes by random forest with 70.5% accuracy. In our random forest model, there are seven explanatory variables: functional enrichment P-value, network topology score, network connectivity, network size, average node degree, unknown gene ratio, and the pattern recurrence numbers.

The predicted gene functions cover a wide range of functional categories, e.g. protein biosynthesis, electron transport, vesiclemediated transport, immune response. Each prediction is made conditionally on the gene's network environment and specific perturbations. Some functions, such as protein biosynthesis, occur universally under almost all perturbations. Others, such as cell cycle, are activated predominately in conditions related to cancer and development. The comprehensive prediction results, together with the necessary context information, are available at http://zhoulab.usc.edu/ContextAnnotation. Many of our predictions are supported by experimental studies in the literature. For example, we predicted NCF4 to participate in "immune response". According to a study [Wientjes et al., 1993], NCF4 is important for immunity and its deficiency leads to chronic granulomatous disease (CGD). We assigned the function "mitotic cell cycle" to AURKB; and AURKB is known to be responsible for mitotic arrest in the absence of aurora A [Yang et al., 2005]. We predicted RPS8 to be involved in "protein biosynthesis", and RPS8 has been shown to participate in translation [Yu et al., 2005]. Spc25 was predicted to be involved in "mitotic cell cycle", which is supported by the evidence that SPC25 is an essential kinetochore component that plays a significant role in proper execution of mitotic events [Bharadwaj et al., 2004].

It should be noted that the prediction accuracy of 70% is an underestimate due to the sparse nature of human GO annotations. Since GO annotation is based only on positive biological evidence, many annotated genes may still have other undiscovered functions. Furthermore, the GO directed acyclic graph structure is not perfect. For example, we predicted ch-TOG to have "mitotic cell cycle" function. Based on recent

132

evidence [Cassimeris and Morabito, 2004], the updated (2006 Dec) GO classifies it as "RNA transport", "centrosome organization and biogenesis", "spindle pole body organization and biogenesis" and "establishment and/or maintenance of microtubule cytoskeleton polarity". Since none of these four GO nodes is a child of "mitotic cell cycle", we have to classify the prediction to be wrong, while the original paper clearly documented its important involvement in mitotic cell cycle [Cassimeris and Morabito, 2004].

6.3.4 Context-Specific Function Annotation

One of the surprises of the human genome project is that we have far fewer genes than expected. A possible explanation to relate the limited number of genes to the high degree of complexity is that many genes perform multiple functions. Since our approach allows one gene to appear in more than one network module, we are able to perform context-sensitive functional annotation. That is, we can assign to a gene multiple functions as well as the network environments in which the gene exerts those functions. This is valuable even if a gene's function is already known. Among our predictions, 20% of genes are assigned multiple functions. This is certainly an underestimate, since for each network module we only picked the functional category associated with the highest network topology score. Of course, some of the different assignments for the same gene are relevant, such as "response to pest, pathogen or parasite" and "immune response", or "regulation of cell cycle" and "mitotic cell cycle". However, the dramatic difference in the network environment associated with those functional assignments indicates different functional involvement of this gene, which is beyond the rough classification of GO functional categories. In our predictions, among the genes assigned with multiple functions, 72% are in network modules that differ at least 50% of member genes and 57% are in network modules that differ at least 70% of member genes.

Figure 6.5 shows two examples of genes predicted to have multiple functions. In Figure 6.5(a), IRF1 appears in three different network modules, and annotated with the functions "immune response", "regulation

133

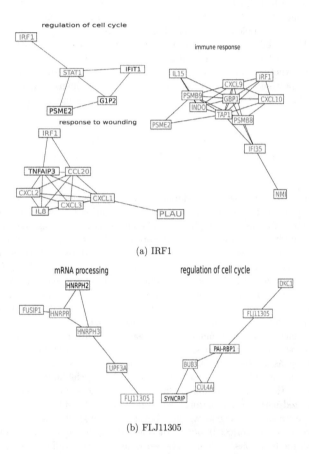

(a) IRF1

(b) FLJ11305

Figure 6.5: Functional predictions for (a) IRF1 and (b) FLJ11305 upon different network environments. Nodes labeled in red are annotated with the titled functions by GO consortium. Nodes labeled in green are the genes with predicted functions.

of cell cycle", and "response to wounding", respectively. The first pattern appears in 6 datasets (details on Supplementary website). The dataset conditions include cancer, infection, and inflammatory responses, which is consistent with IRF1's role in "immune response". The second pattern appears also in 6 data sets, measuring exercise effect, infections, and cancer. Since cell cycle may also be accelerated upon inflammatory responses, and conditions such as "cancer" may impact various pathways, it is hard to conceptually separate those two datasets into two types of strictly different conditions. In fact, in two of those datasets measuring infection (GDS260 and SMD dataset with Category=Infection, Subcategory=PBMC, experimenter=Cheryl Hemingway), the two network modules merge into one, indicating a potential role of IRF1 in mediating cross-pathway communication. The third pattern annotated with the function "response to wounding" occurs in 5 datasets. Since the process "response to wounding" is highly related to the process "immune response", and it may also initiate the acceleration of "cell cycle", the dataset conditions are similar to those previously described, except the condition "osmotic stress reaction", which is in agreement with the specific process "response to wounding". The first two functions "immune response" and "regulation of cell cycle" agree with the known annotation of IRF1, and we believe that the function "response to wounding" is also likely to be true due to strong evidence (The hyper-geometric P-value measuring the module functional homogeneity is 10^{-5}). As another example, Figure 6.5(b) shows that the unknown gene FLJ11305 occurs in two different network modules, and is annotated with two functions "mRNA processing" and "regulation of cell cycle". Each module is activated in 5 datasets, including drug treatment, dyslipidemia, Huntingtons disease, exercise effect, and cancer, with the conditions dyslipidemia and ovarian tumor being shared between the two modules. The fact, that often different network modules involving the same gene can be merged together under some conditions, indicates that many genes with multiple functions may participate in related pathways, and they are likely to serve as cellular process communicator.

135

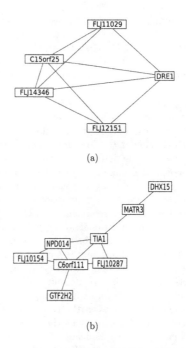

(a)

(b)

Figure 6.6: Two examples of uncharacterized Cellular Systems.

6.3.5 Discovery of Uncharacterized Cellular Systems

The comprehensive functional modules generated in this study can facilitate the discovery of uncharacterized cellular systems. Date and Marcotte defined "uncharacterized cellular systems" as discrete subgraph in reconstructed protein interaction networks in which 50% or more member proteins lack functional assignments [Date and Marcotte, 2003]. Among the identified modules, we identified 2206 such modules, varying in size from 4 to 68 member genes. Among those, 204 modules contain only unannotated genes.

Figure 6.6(a) shows an example. The module contains 5 genes, DRE1, C15orf25, FLJ11029, FLJ12151, and FLJ14346, that form a densely connected subgraph. The complete subgraph appears in 8 datasets (GDS1062, 1312, 1321, 505, 564, 760, 858, 914). Notably, 5 out of the 8 data sets are cancer data sets. Although cancer data sets are enriched in our collected data (21 out of 65), the ratio is still marginally significant at ($P = 0.06$) level. This suggests the potential involvement of the module in cancer. Interestingly, the homolog of DRE1 in Drosophila plays an important role in regulating DNA replication-related genes [Okudaira et al., 2005], which may suggest its potential role in cell cycle or cell proliferation — a hypothesis consistent with its activation in cancer. This example demonstrates that even for the poorly uncharacterized modules, our method may provide useful information based on the experimental conditions under which the module is activated.

Furthermore, 251 uncharacterized modules have connectivity less than 0.5, which can hardly be identified from a single graph. Figure 6.6(b) shows such an example. Among the eight member genes, three have annotated functions: DHX15 is involved in mRNA processing, GTF2H2 participate in regulation of transcription, and TIA1 is a member of a RNA-binding protein family. The exact function of MATR3 is unknown, but it is known to encode a nuclear matrix protein, which may play a role in transcription. These evidences point to a possibly involvement of the module in transcription.

6.4   Conclusion and Discussion

We have presented a generic approach to integrate many microarray datasets to identify functional modules and to perform functional annotation in human genome. To our knowledge, this is the first study to systematically annotate human gene functions based on multiple microarray datasets. Compared to current approaches based only on a single microarray dataset, our method provides: (1) higher specificity: the identified functional modules are more likely to be functionally homogenous; (2) higher sensitivity: we can identify functional modules beyond co-expression clusters. Our approach is based on pattern mining across co-expression networks. It is known that absolute expression values of a gene cannot be compared across data sets. However, the expression correlations of a gene pair in different datasets are comparable because they are unitless measures each derived from a single data set. As our co-expression networks are constructed from expression correlations of gene pairs, their comparisons are not affected by inter-dataset variations. Thus, our approach provides an effective way to integrate a large number of microarray experiments conducted in different laboratories, at different times, and using different technology platforms. There are large numbers of public microarray data sets available for model organisms such as H. sapiens, M. musculus, A. thaliana, D. melanogaster, C. elegans, and S. cerevesiae. Using our approach, we are in a position to extract orders of magnitude more information for any genome, for which large amount of microarray data exists. A natural extension will be to compare co-expression networks across species. Several studies along this direction [Bergmann et al., 2004, Oldham et al., 2006, Stuart et al., 2003] have already been performed on two or several species. To extend those studies to many species is likely to require efficient algorithm design.

In our studies, 20% of genes received multiple functions in different network contexts. This is certainly an underestimate for at least three reasons: (1) we only assigned one function to a gene based on its top network topology score; (2) the whole functional module may perform multiple functions in different contexts of activities of other modules; (3) our data source only includes 65 datasets, that mostly represent human pathological conditions and cover a small proportion of human dynamical functional landscape. We note that incorporating the concept of dynamics is especially important in charactering human gene functions due to its high temporal and spatial complexity. However, relating specific conditions to particular gene functions is not an easy task due to the subtle difference in experimental conditions, and the difficulties in systematically characterize them.

The principle of our approach, integrating multiple networks for functional studies, can be extended beyond microarray analysis. For example, a popular approach in identify functional modules is to identifying dense subgraphs on protein interaction networks [Chen and Yuan, 2006, Hwang et al., 2006, Koyuturk et al., 2006, Luo et al., 2007, Spirin and Mirny, 2003, Tornow and Mewes, 2003]. However, as discussed above, functional modules often occur as non-dense subgraphs, e.g. metabolic and signal pathways. Furthermore, since current protein interaction network are static networks, edges in such a network may not occur together if considering temporal or spatial parameters. Thus, such identified functional modules may not truly represent a functional unit. In the future, given protein interaction networks generated under different conditions, our approach can further facilitate the identification of condition-specific functional modules or dynamic protein complex assembly. In fact, if different species are conceptualized to represent manifestations of different conditions of life forms, several recent studies on conservation and evolution of protein interaction networks across species can be regarded as a first attempt to characterize network dynamics [Flannick et al., 2006, Kelley et al., 2003, Koyuturk et al., 2006, Sharan and Ideker, 2006, Sharan et al., 2005]. In that context, due to the NP-hard graph isomorphism problem, how to perform large-scale pattern mining across protein interaction networks of many species is still a challenging problem.

Chapter 7

A Quantile Method for Sizing Optical Maps

7.1 Introduction

Optical mapping is a proven, high-throughput, single-molecule system that constructs physical maps spanning entire genomes. Because restriction maps describe large-scale structure (0.5 Kb — entire chromosomes), they intrinsically reveal a broad range of genomic alterations reflecting polymorphisms and aberrations. Also, such maps aid genome sequencing projects by facilitating assembly, independent validation and finishing efforts [Zhou et al., 2006]. Although optical mapping analyzes clones and PCR amplicons, significant automation efforts have centered on the construction of whole genome maps using large, randomly sheared genomic DNA molecules [Armbrust et al., 2004, Reslewic et al., 2005]. Here, a microfluidic device deposits genomic DNA molecules as a series of centimeter long stripes onto critically charged glass surfaces. Since large molecules exist in solution as random coils, this procedure unravels them, producing a stretched form, which is electrostatically adhered to the surface, thus ensuring optimum presentation prior to restriction digestion. After staining with a fluorochrome dye, these cleavage sites are imaged by fluorescence microscopy as micron-sized gaps formed due to the local

141

release of stored tension within stretched vs. relaxed molecules. Restriction fragments retain their order and an automated image-acquisition and machine-vision system provides capacious single molecule data sets for analysis. Optical map construction uses integrated fluorescence intensity measurements for estimation of restriction fragment sizes. This procedure considers internal fluorescence standards consisting of co-mounted small DNA molecules (reference molecules) of known sequence composition for establishing a standard measure of integrated fluorescence intensity per Kb of DNA; a simple relationship then sizes restriction fragments [Lin et al., 1999] as follows:

$$\text{Estimated fragment size (kb)} = \frac{\text{integrated intensity for fragment}}{\text{standard intensity per Kb}}$$

$$(7.1)$$

This sizing method suffers errors in the estimated standard intensity because of the estimation in Equation (7.1). Such errors arise because optical mapping estimates restriction fragment sizes by essentially assessing the ratio of dye molecules bound to a target and reference, and these measurements must deal with issues regarding dye uptake by DNA molecules bound to surfaces [Dimalanta et al., 2004].

We overcome these limitations by development of a new approach for the estimation of restriction fragments associated with optical maps. Currently our approach assumes that a reference genome is available. We treat the sizing problem as a "blind inversion" problem. That is, we regard the process of measuring the fluorescence signals from single molecules as a system that considers inputs and outputs as just fragment sizes and intensities. Although both the effective system and input DNA fragment sizes are unknown to us, we can still estimate sizes given their intensities through a quantile-quantile correspondence. The procedure is dictated by the rationale of Blind Inversion Needs Distribution (BIND) [Li, 2003]. A similar quantile method has been applied to the normalization of microarrays [Bolstad et al., 2003]. Since our method uses only fluorescence intensity determinations of analyte molecules and a sequence from a reference genome, this approach eliminates the need for assaying internal reference molecules, and renders moot any errors stemming from such measurements. Furthermore, reduction of experimental variables gives increased simplicity, and this can foster development of new single molecule platforms aimed at genomic analysis. Here, prelimi-

142

nary analysis shows that the performance of our approach is comparable to that of the current reference molecule method as judged by rate of map placements to a reference map.

The remainder of the chapter is organized as follows. Section 7.2 describes the proposed method with discussion regarding the treatment of typical errors associated with optical maps, including: missing or spurious (false) restriction sites and missing fragments. In Section 7.3, we describe experimental results on some optical mapping data sets. Section 7.4 concludes the chapter with some directions of further research.

7.2   Method

A statistical model for optical mapping.   Our method utilizes the following statistical models dealing with measurement errors associated with optical mapping described in [Valouev et al., 2006], and we briefly summarize them in what follows.

- **M1. A Poisson model for enzyme cuts.** Sizes (in Kb) Y of genomic restriction fragments have an exponential density with mean $EY = \lambda$ that depends on the endonuclease used for the digestion. Consequently, the number of restriction sites in s Kb of DNA is a Poisson process with intensity s/λ [Churchill et al., 1989, Waterman, 1995]. Typical enzymes produce fragments with average lengths $12 - 40$ Kb.

- **M2.   Missing cuts.** After the DNA is enzymatically cleaved, some restriction sites are left uncut by the endonuclease (partial digestion). Observations of restriction sites on the optical maps are assumed to be Bernoulli trials with the digestion rate p (usually $p \approx$ 0.8). Hence observed restriction sites are explained by a thinned Poisson process with intensity p/λ [Grimmett and Stirzaker, 1982].

- **M3.   False cuts.** False cuts result from random DNA breaks or non-specific activity of endonuclease. We assume that random breaks show no sequence dependent bias and are uniformly distributed across the entire reference genome. According to the model, the number of false cuts per s Kb of DNA is a Poisson process with intensity ζs. On average, we observe about five false cuts per 1 Mb of DNA.

- **M4. Measurement error.** During the staining of DNA, the fluorescent dye randomly binds to sites on DNA molecules, despite saturation conditions. For a true fragment size Y, its observed fluorescence intensity X approximately follows a normal density $N(\mu Y, \mu \sigma^2 Y)$ for sufficiently large fragments (e.g., $> 4\,Kb$). This result is analytically supported by the central limit theorem [Valouev et al., 2006].

In addition, we have the following considerations.

- **Missing fragments.** Because electrostatic forces are used for retaining DNA molecules, smaller restriction fragments preferentially detach from the surface, and are consequently under-represented in data sets.

- **Chimeric maps.** Molecules that physically overlap on glass surfaces cannot be spatially resolved for revealing true connectivity of their respective portions. Consequently, some maps are chimeric, meaning that they represent multiple unrelated genomic regions.

Distribution of observed fragment intensities and an exponential approximation Under the model assumptions, it was shown in [Valouev et al., 2006] that measured sizes X of optical map fragments have an exponential density with the mean

$$\theta = \mu \left[\frac{1}{\sigma} \sqrt{\frac{2}{\tau} + \frac{1}{\sigma^2}} - \frac{1}{\sigma^2} \right]^{-1} \qquad (7.2)$$

for $X \geq \Delta$, where Δ is a positive number and $\tau = \left(\zeta + \frac{p}{\lambda} \right)^{-1}$. We note that the exponential distribution can be viewed as an approximation to that of the real fragment intensities. In Figure 7.1 we plot the histograms

144

of intensities from a human mapping project (see Section 7.3) using the enzyme SwaI with and without chimeric reads. Note that a large fraction of short fragments are missing as can be seen at the lower ends of the histograms. Also, histograms have longer tails than an exponential distribution. In the Q-Q plot, data are from samples in the quantile interval $(0.15, 0.95)$ and the reference exponential distribution has mean 1.27×10^6. It is clear that the distribution of the real fragment intensities is well approximated by an exponential distribution except for the lower and upper tails.

Distribution of fragment lengths Now let us consider the hypothetical situation in which one fluorescence unit corresponds to exactly one base pair. If we apply the exponential approximation (7.2), then we have $\mu = 1$. Namely, measured fragment lengths of optical maps follow an exponential density with the mean $\theta = \left[\frac{1}{\sigma} \sqrt{\frac{2}{\tau} + \frac{1}{\sigma^2}} - \frac{1}{\sigma^2} \right]^{-1}$. Note that the three parameters, the digestion rate p, the false break rate ζ, and the standard deviation of measurement σ are confounded in the expression of the exponential scale parameter. To reflect the detailed features of the human genome, we can simulate missing cuts and false cuts on the human genome according to M2 and M3 and common rate values. That is, restriction sites are skipped by 20% independently of each other, and on average 5 false cuts per 1 Mb are induced according to a Poisson process. For the moment, we ignore the measurement error in the simulation. The distribution of the fragment lengths is shown in Figure 7.2(a). Although a longer tail exists on the right, the middle part of the histogram looks fairly close to an exponential density. This is confirmed by the Q-Q plot of data versus exponential in the quantile range $(0.15, 0.95)$. Furthermore, we observe excessive short fragments under 1 Kb in the log-scale Q-Q plot. The deviations from the exponential distribution on the lower and upper ends indicate the non-randomness of DNA sequences. Although it is well known that genome sequences are non-random, Churchill *et al.* found that the exponential density fits best for modeling the distribution of restriction fragment sizes compared to several other densities they examined in the case of *E. coli* [Churchill et al., 1989]. This assumption remains largely valid for other genomes from our experience. However, the assumption may be violated at the lower and upper ends as shown in Figure 7.2(a) and 7.2(b). We take into

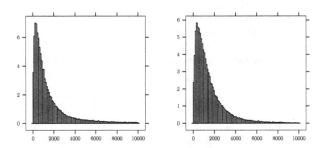

(a) Histogram of observed fragment in- (b) Histogram without chimeric reads
tensities with chimeric reads

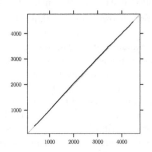

(c) Q-Q plot without chimeric reads

Figure 7.1: Histograms and a Q-Q plot from SwaI-human optical maps with and without chimeric reads. In figures, intensity unit is 1000. Histograms have longer tails than an exponential distribution. In the Q-Q plot, data are from samples in the quantile interval $(0.15, 0.95)$. Corresponding exponential distribution has mean 1.27×10^6.

account these issues in our sizing scheme by a censoring strategy.

Correspondence of the two distributions Suppose we have n observed intensities $x_1, \ldots, x_n$. Their corresponding fragment lengths are denoted by $y_1, \ldots, y_n$, which are unknown. The goal is to estimate fragment length y_i of given intensity x_i. We formulate their relationship by

$$x = h(y) \qquad (7.3)$$

where $h(\cdot)$ represents the system of optical mapping instruments. In this formulation, both input y and the system function h are unknown. This is a typical "blind inversion" problem that appears in many scientific measurement problems such as DNA sequencing [Li, 2003]. If it is true that larger fragment sizes have larger intensities, then we assume that
A1. h is monotone. That is, if $y_i \leq y_j$, then $x_i \leq x_j$.
Sort respectively the intensities and fragment sizes in the ascending orders, namely, $y_{(1)} \leq y_{(2)} \leq \cdots \leq y_{(n)}$, and $x_{(1)} \leq x_{(2)} \leq \cdots \leq x_{(n)}$. Under the monotone transformation, the size of $y_{(i)}$ should map to $x_{(i)}$. But the latter are unfortunately not observed. However, if the distribution of the input is known, we can approximate $\{x_{(i)}\}$ by the i/n-th quantiles of x. Ideally, the larger the n is, the more accurate the approximation is. Explicitly, the rationale for the sizing method is:
A2. The distribution of the reconstructed fragment lengths should match the expected distribution of fragment lengths.
If we have the complete data set of all fragment intensities, small or large, the problem is solved since we can obtain the expected distribution of fragment lengths as illustrated earlier. In this case, we even do not need to assume any parametric model other than the random mechanisms of missing and false cuts. However, the size estimation is difficult because many short fragments are missing in the observed intensities as shown in Figures 7.1(b) and 7.1(c). Besides, there are only a few very large fragments and the match of quantiles on the very right end is difficult. To be able to apply our method, we use the following strategy to deal with the complications.
A3. For both the distribution of the observed fragment intensities and the expected distribution of fragment lengths, the middle part of the distribution approximately follows a truncated (from both ends) exponential distribution.

147

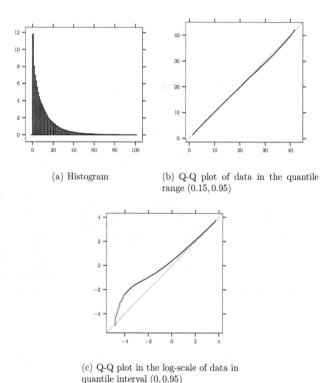

(a) Histogram

(b) Q-Q plot of data in the quantile range $(0.15, 0.95)$

(c) Q-Q plot in the log-scale of data in quantile interval $(0, 0.95)$

Figure 7.2: Histogram and Q-Q plots of fragment sizes (Kb) of human restriction map for enzyme SwaI. The histogram has a longer right tail (not shown in the figure) than the exponential distribution. Note that some long fragments could be more than 400kb. In the Q-Q plots, the horizontal axis is the sorted fragment sizes of human restriction map while vertical axis is the quantiles of a reference exponential distribution with mean 11.49. Clearly, the major part of data follow an exponential distribution well as verified in (a) and (b). However, (c) indicates that there exist excessive short fragments (under 1kb).

Figures 7.1 and 7.2 show supporting evidence to this claim. Moreover, we describe each of the two distributions respectively by a mixture model of three components. The principal component is an exponential distribution; two other components respectively have their mass located mostly at the lower end and the upper end. Note that the mixing fractions of these two components are small and we avoid assuming any specific forms. Besides, a fraction of small fragments is missing in the distribution of intensities. To deal with the difficulties on the two ends, we modify the scheme as follows: first we estimate the fragment sizes in the middle part by matching quantiles; second, we estimate the fragment sizes on the two ends by the ratio of the two exponential scales, respectively estimated from intensities and fragment lengths of the simulated reference genome.

MLE for the exponential scale and two cutoff quantiles from the truncated data As explained earlier, the major and middle part of both intensities and fragment sizes can be approximated by exponential distributions. To reduce the influence of the two end components on the parameter estimation of the principal exponential distribution, we remove data from the two ends. The above sizing schemes requires estimates of the parameters in the exponential distribution. They include the scale parameter and the two quantiles corresponding to two given cutoff values.

We propose two algorithms to compute the parameter estimates. One assumes that the middle part of data follows a doubly truncated exponential distribution and uses standard maximum likelihood estimation (MLE) method to estimate its mean. The other approach is an EM algorithm [Dempster et al., 1977, Redner and Walker, 1984], which treats the exponential components on the two ends as missing data. Our simulation (not shown) indicates that both algorithms perform well and give consistent estimates.

Let X be a random variable following an exponential distribution with mean λ. Suppose we have n observations $x_1, \ldots, x_n$. Given two cutoff values $a < b$, we remove observations such that $x_i < a$ or $x_i > b$ and treat them as missing data. Suppose n' observations are left and are denoted by $x'_1, \ldots, x'_{n'}$. Now the remaining samples $x'_1, \ldots, x'_{n'}$ are

149

drawn from a doubly truncated exponential distribution [Cohen, 1991] with density

$$f(x; \lambda) = \begin{cases} 0 & x < a \text{ or } x > b \\ \frac{\frac{1}{\lambda}e^{-x/\lambda}}{F(b)-F(a)} & a \leq x \leq b \end{cases} \qquad (7.4)$$

where $F(x) = 1 - e^{-x/\lambda}$ is the distribution function of exponential distribution. Therefore, the log likelihood function is

$$\mathcal{L} = -\frac{1}{\lambda} \sum_{i=1}^{n'} x_i' - n' \log[\lambda[F(b) - F(a)]] \qquad (7.5)$$

Newton-Raphson algorithm   By setting the derivative $\partial\mathcal{L}/\partial\lambda$ equal to zero, we have

$$\lambda = \bar{x} - a + \frac{(b-a)[1 - F(b)]}{F(b) - F(a)} \qquad (7.6)$$

where $\bar{x} = \frac{1}{n'}\sum_{i=1}^{n'} x_i'$. Because no closed form solution exists, we compute a numerical solution using the Newton-Raphson algorithm.

EM algorithm   Starting with an initial value of λ, iterate over the following steps until λ_k converges.

E-Step   For the current parameter value λ_k, calculate $q_a = \Pr(X < a|\lambda_k)$ and $q_b = \Pr(X < b|\lambda_k)$. Since n' observations are in the range $[a, b]$, on average $n'q_a/(q_b - q_a)$ and $n'(1 - q_b)/(q_b - q_a)$ samples are missing in the lower and upper ends, respectively. We impute the value of the lower end samples by $x_a = E(X|X < a; \lambda_k) = (\lambda_k - (a + \lambda_k)e^{-a/\lambda_k})/(1 - e^{-a/\lambda_k})$ and the upper end ones by $x_b = E(X|X > b; \lambda_k) = b + \lambda_k$.

M-Step   Combine observations and imputed missing data, and compute the standard MLE on the complete data. Namely,

$$\lambda_{k+1} = \frac{\sum_{i=1}^{n'} x_i' + x_a n'q_a/(q_b - q_a) + x_b n'(1 - q_b)/(q_b - q_a)}{n'/(q_b - q_a)} \qquad (7.7)$$

Detailed sizing scheme. Given an optical map data of n intensities, we sort them in the ascending order. We choose two cutoff values a and b corresponding respectively to two quantiles of the raw intensities, r_a and r_b such that $0 < r_a < r_b < 1$. Suppose n_1 intensities are smaller than a, n_2 intensities are larger than b, and n' intensities are in $[a, b]$, where $n_1 + n' + n_2 = n$. We next apply either the proposed EM or Newton-Raphson algorithm to the n' intensities in $[a, b]$ to estimate the mean value λ of underlying exponential distribution. Let $q_a = F(a|\lambda)$ and $q_b = F(b|\lambda)$. Similar to the imputation in the EM algorithm, there should be, on average, $n_a = q_a n' / (q_b - q_a)$ intensities less than a and $n_b = (1 - q_b) n' / (q_b - q_a)$ intensities greater than b. Note that the imputation is for the principal exponential component. Denote the sorted intensities in the range $[a, b]$ by $a \le y_{(n_1+1)} \le y_{(n_1+2)} \le \dots < y_{(n_1+i)} \le \dots y_{(n_1+n')} \le b$. As a result, the quantile of $y_{(n_1+i)}$ in the principal exponential distribution is given by

$$q_i = \frac{i + n_a}{n' + n_a + n_b} \tag{7.8}$$

Similarly, we apply the above procedure to fragment sizes of the reference map to obtain the adjusted quantiles excluding excessive short and long fragments. Denote by λ_r the estimated mean of fragment sizes computed from the EM or Newton-Raphson algorithm.

Finally, we can map intensities $\{y_i\}$ in $[a, b]$ to fragment sizes $\{x_j\}$ at the same quantile. For intensities less than a or greater than b, we cannot estimate their accurate quantiles because they do not follow the same distribution. Instead, we estimate the standard intensity μ by λ/λ_r and determine the fragment size x of an intensity y by $x = y/\mu$ if $y < a$ or $y > b$. Note that the fragment sizes estimated in this way may differ from those obtained through quantile mapping at points a and b. Thus, we make the following continuity correction: shift the sizes on each end by a constant so that the estimated fragment sizes are continuous at cutoff points a and b. We summarize the entire sizing procedure in Figure 7.3.

Practical issues Clearly, q_a and q_b are critical for the precision of size estimation. The interval (q_a, q_b) should be appropriately selected so that we use as many fragments as possible in the quantile method

ALGORITHM A QUANTILE METHOD FOR SIZING OPTICAL MAPPING FRAGMENTS

Input: Intensities $y_1, \ldots, y_n$ of an optical map, the reference genome sequences and a given restriction enzyme, digestion rate p, random breakage rate ζ, and parameters r_a and r_b such that $0 < r_a < r_b < 1$.
Output: Fragment sizes.
Steps:

1. Generate reference restriction map of given DNA sequence and enzyme. Simulate missing cuts and false cuts with digestion rate p and random breakage rate ζ on the whole sequence, respectively. Repeat the simulation sufficient times and denote simulated fragment sizes $x_1, \ldots, x_m$.

2. Sort intensities. Let a and b be intensities at quantiles r_a and r_b, respectively. Suppose n_1 intensities are smaller than a and n' intensities are in $[a, b]$. Apply the proposed EM or MLE algorithm to intensities such that $a \leq y_i \leq b$. Denote the estimated mean of exponential distribution by λ.

3. Let $q_a = F(a|\lambda)$, $q_b = F(b|\lambda)$, $n_a = q_a n'/(q_b - q_a)$, and $n_b = (1 - q_b)n'/(q_b - q_a)$. For intensities $a \leq y_{(n_1+1)} \leq y_{(n_1+2)} \leq \cdots < y_{(n_1+i)} \leq \cdots y_{(n_1+n')} \leq b$, calculate the corresponding quantiles of $y_{(n_1+i)}$ by $q_i = \frac{i+n_a}{n'+n_a+n_b}$.

4. Repeat Steps 2 and 3 on simulated fragment sizes. Denote the estimated mean of the underlying exponential distribution by λ_r.

5. For intensities $a \leq y_i \leq b$, map to the fragment sizes at the same quantiles.

6. For intensities $y_i < a$ or $y_i > b$, calculate the corresponding size as $y_i \lambda_r/\lambda$ and shift them so that the estimated sizes are continuous at the cutoffs.

Figure 7.3: The algorithm of the quantile sizing method.

whereas excessive short and long fragments are excluded. Histograms and Q-Q plots are helpful for selecting q_a and q_b. We may also optimize them according to the placement performance [Valouev et al., 2006]. The digestion rate p and random breakage rate ζ can be estimated from abundant previous optical map data by fitting them to the reference map.

7.3 Method Validation

Experiments We implemented our method by analysis of optical mapping results from four separate "mounts" that were part of an ongoing human genome mapping project. The DNA was digested using the restriction enzyme SwaI. Mounts 1 and 2 are of high quality while 3 and 4 are of typical quality.

The quality of these measurements is assessed by the number of "flagged" restriction fragments assigned by the image processing software. Briefly, the software "flags" an imaged fragment when associated pixels show abnormally high fluorescence intensities (Runnheim, private communication). This phenomenon occurs — as one example — when molecules overlap, giving rise to chimeric maps. Because such occurrences produce ambiguous results, we remove these fragments from the quantile sizing procedure.

Typically, the optical mapping microfluidic device deposits 48 independent stripes (channels) of DNA molecules onto a surface. However, measurements from only 47 and 27 channels were acquired for Surfaces 1 and 4. The details of the four data sets are described in Table 7.1. Surface 1 has 77986 fragments from 4370 molecules, of which 22.9% fragments are flagged. Surface 2 has 75991 fragments from 4565 molecules, of which 22.2% fragments are flagged. Surface 3 has 73843 fragments from 5014 molecules, of which 28.6% fragments are flagged. Surface 4 has only 48596 fragments from 3015 molecules since only 27 channels are available. Among them, 26.3% fragments are flagged. Clearly, Surfaces 1 and 2 are of better quality than Surfaces 3 and 4 in terms of percentage of flagged fragments.

Table 7.1: Description of the test data sets.

Surface	Channels	Fragments	Flagged fragments	Molecules	Molecules with ≥ 20 non-flagged fragments
1	47	77986	22.9%	4370	852
2	48	75991	22.2%	4565	769
3	48	73843	28.6%	5014	437
4	27	48596	26.3%	3015	357

Given that this data set consists of randomly sheared molecules from a human genome, we do not have independent validation of our size measurements. As such, we take an alignment strategy to indirectly assess the sizing scheme. That is, we align optical maps sized by both the reference-DNA and the quantile methods to the reference genome, and then check the number of significant and consistent placements. The genomic identity of a single molecule optical map is found through its alignment to a reference genome. When fragment sizes are precisely estimated, the alignment procedure correctly places a map; this is noted by a significant score. However, when maps show a small number of fragments, often they are ambiguously placed at many genomic locations. Accordingly, we only consider maps containing at least 20 non-flagged restriction fragments.

The alignment and scoring system   We use the dynamic programming algorithm in [Waterman et al., 1984, Valouev et al., 2006] to compute the optimal restriction map alignments. Define a reference map $R = (r_0, \ldots, r_m)$ by a set of site positions relative to the start of the map, i.e. $0 = r_0 < r_1 < \ldots < r_m$. Similarly, let $Q = (q_0, \ldots, q_n)$ be an optical map such that $0 = q_0 < q_1 < \ldots < q_m$. The alignment between R and Q is given by a sequence of ordered pairs of matching site indices $(i_0, j_0), (i_1, j_1), \ldots, (i_d, j_d)$, such that $i_0 < i_1 < \ldots < i_d$, $j_0 < j_1 < \ldots < j_d$, i_t and j_t correspond to sites r_{i_t} and q_{j_t} of maps R and Q. Let $S(i, j)$ be the optimal score of the alignment between R and Q that the rightmost pair of sites i and j are aligned. The algorithm calculates the score $S(i, j)$ using the recursion

$$S(i, j) = \max_{(0 \vee i - \delta_1) \leq g < i, (0 \vee j - \delta_2) \leq h < j} [S(g, h) + X(q_i - q_g, r_j - r_h, i - g, j - h)]$$

(7.9)

where δ_1 and δ_2 are parameters that specify the sizes of maximum matching regions respectively for the reference and optical maps, and the function $X(q_i - q_g, r_j - r_h, i - g, j - h)$ calculates the score for the region $(g, h) \rightarrow (i, j)$.

The alignment algorithm adopts the scoring function proposed by Valouev *et al.* [Valouev et al., 2006]. The scores are log-likelihood ratios under two hypotheses: a given optical map is indeed from the region to which it is aligned or is independent of the region. The odds are based on the statistical model of optical map measurements and other complications such as missing cuts and false cuts, see M1-4 or [Valouev et al., 2006] for details. Currently we treat the flagged fragments as missing when fitting optical maps to reference map. The scores for flagged fragments are simply set to be zero at this point.

Results The fragment intensities of the four data sets are converted into fragment sizes by both the reference DNA and quantile methods. The obtained optical maps are then fitted to the reference genome. We want to compare the two sizing methods. In the quantile sizing method, the values of the two end quantiles r_a and r_b are around 0.30 and 0.95 respectively. In the scoring function of the alignment algorithm, the parameter λ, the mean of reference exponential distribution, is set to be 12.72 Kb, the average fragment length of the reference map; the maximum matching region size δ is 5; the standard deviation σ of size error for long fragments ($\geq 4kb$) is 0.46; the standard deviation η of size error for short fragments ($< 4kb$) is 0.53. The details of these parameters can be found in [Valouev et al., 2006]. The values of the digestion rate and the random breakage rate are typical for many optical maps. The parameters r_a, r_b, σ, and η are obtained by optimizing the fitting of long molecules with no flagged fragments to the reference genome.

After obtaining the optimal fitting scores of molecules to the reference genome, we need to determine the significant hits. Note that molecules

Table 7.2: A comparison of the two sizing methods. For each surface, we fit optical maps sized by both methods. A placement is called by two cutoff values 2.0 and 2.3. The numbers of placements by the reference molecule method, the quantile method, and by both methods are shown.

Cutoff value	2.0			2.3		
Sizing method	reference molecule	quantile	both	reference molecule	quantile	both
Surface 1	282	173	133	159	86	67
Surface 2	299	260	223	191	165	136
Surface 3	32	22	11	14	7	7
Surface 4	27	36	17	9	10	7

consisting of more fragments tend to have larger scores. We address this by considering the normalized score S/d, where S and d are respectively the fitting score of a molecule and the number of matching blocks. A matching block is defined to be a minimal set of fragments flanked by matching sites in the corresponding alignment to the genome. The significant placements are then determined based on the normalized scores. In Table 7.2, we list the numbers of significant placements for the cutoff value of 2.0 and 2.3. In order to assess the false positive rate, we did a simulation study. Namely, we simulated random maps and real maps from the reference genome and fit them back to the reference genome by the same set of parameters. The result shows a 5% false positive rate at the cutoff value of 2.3. However, in this simulation, the sizing errors follow the model [Valouev et al., 2006] and do not include the issue of flagged fragments. Thus this evaluation only serves as a guide. We note that more multiple hits are observed in the quantile method.

In Table 7.2, we show not only the numbers of molecules placed by each method but also the numbers of molecules placed by both methods. In Surfaces 2, 3, and 4, large fractions of the significant placements are common to both methods. The consistency for Surface 1 is not as good as others. We checked closely into the placements of the "good" molecules that do not have flagged fragments by examining their alignments. Since false cuts are rare compared to missing cuts, we take 1 for the value of δ_1 and 5 for δ_2. Again most of the significant hits are common to both methods. We checked those "good" molecules placed by the reference

DNA method but not the quantile method, and found they contain more false cuts, indicating the possibility of spurious alignments. There are three molecules placed by the quantile method but not by the reference DNA method.

The average fragment size of the reference genome cut by SwaI is 12.72 Kb. After applying the 20% missing cuts and 5 false cuts per Mb, the average size of fragment lengths becomes 14.76 Kb. As we have explained, the principal exponential component of fragment lengths plays an important role in our approach. If we take 30% and 95% as the truncation thresholds, the estimates of the principal exponential components of the two kinds of fragment lengths are respectively 11.82, 13.85. The estimates of the exponential scale parameter for the sizes determined by the quantile method are 13.08 Kb, 14.80 Kb, 14.91 Kb, 14.92 Kb for Surface 1, 2, 3, 4. In comparison, the corresponding estimates for the sizes determined by the reference molecule method are 15.64 Kb, 18.63 Kb, 18.13 Kb, 17.57 Kb. The difference may be accounted by the sampling bias due to the missing fragments and the flagged fragments. Remember that we do not include the flagged fragments in our quantile procedure. The average sizes of all fragments are larger than the estimates of the exponential scale parameter for both methods and in all chips. This is not a surprise because we only use an exponential density to approximate the principal component.

In Figure 7.4 we show two examples of alignment. The first one is hit by both sizing methods. "Flagged" fragments are represented by "nan". In the second alignment, only the score obtained by the quantile method is significant.

7.4 Discussion

Our quantile sizing method is a robust procedure. In the statistical literature, there are several perspectives of robustness, [Hampel et al., 1986, Huber, 1981]. First, the influence curve or sensitivity curve of the quantiles less than a given $\alpha (< 1)$ is bounded. That is, the influence of

(a) One alignment hit by both sizing methods. "Flagged" fragments are represented by "nan".

(b) One alignment hit by only the quantile method

Figure 7.4: Two alignment examples.

158

an outlier of any kind is finite. As α approaches 1, the sensitivity gets higher as the density function decreases to zero. This is one reason why we exclude the very large fragments in the quantile matching. Second, the breakdown value of the α-th quantile is $\min(\alpha, 1 - \alpha)$. This says that only quantiles on the two extremes are sensitive to perturbations. In the proposed approach, we re-scale fragments on the two ends by extrapolating size information in the middle part. Our numerical results support the view that our method is robust.

Even though the major and middle part of the observed fluorescence intensities is well approximated by an exponential distribution, we do not use exponential quantiles for size estimation. Rather, we use the quantiles obtained by mimicking the missing and false breaks on the reference genome. Importantly, the exponential model is only used to determine the quantiles of the two cutoffs.

Here, the quantile method has only been applied for the analysis of molecular fluorescence intensities derived from each surface. Thus we have ignored any potential spatial effect. Recall that an optical mapping microfluidic device contains multiple channels. A natural extension is to apply the method to each channel. But in this data set, a single channel contains about 100 molecules and may not be sufficient to accurately estimate the sizes. Alternatively, we can divide the surface into subareas and apply the quantile method to each subarea.

According to our preliminary tests, the quantile method is quite consistent with the current reference DNA method. Since the two perspectives are entirely independent, the new method can help validate the results obtained from the reference DNA method. In addition, it could be very important in situations where placement of reference DNA is difficult.

We describe a quantile method of sizing in the context of fitting optical maps to a reference genome. The same idea with necessary modifications may be applicable to the optical map assembly problem. In this problem, all pairwise overlaps are computed between optical maps [Valouev et al., 2006] and the reference genome is unknown. A possible scheme for size estimation is sketched as follows. First we compare the distribution of fragment intensities with an exponential distribution using Q-Q plots. Second, we map intensities to a reference exponential distribution instead of a known reference genome. The mean λ of the reference distribution can be guessed from a highly related genome or simply a random genome. We use the exponential distribution of mean $\tau = (\zeta + p/\lambda)^{-1}$ to simulate fragment sizes. Third, we fit an exponential distribution to the observed intensities excluding the lower and upper ends and apply our method to estimate the sizes of optical maps. We will explore this scheme in our future research.

Chapter 8

The Regularized EM Algorithm

8.1   Introduction

In statistics and many related fields, the method of maximum likelihood is widely used to estimate an unobservable population parameter that maximizes the log-likelihood function

$$L(\Theta; \mathcal{X}) = \sum_{i=1}^{n} \log p(x_i|\Theta) \tag{8.1}$$

where the observations $\mathcal{X} = \{x_i | i = 1, \ldots, n\}$ are independently drawn from the distribution $p(x)$ parameterized by Θ. The Expectation Maximization (EM) algorithm is a general approach to iteratively compute the maximum-likelihood estimates when the observations can be viewed as *incomplete data* [Dempster et al., 1977]. It has been found in most instances that the EM algorithm has the advantage of reliable global convergence, low cost per iteration, economy of storage, and ease of programming [Redner and Walker, 1984]. The EM algorithm has been employed to solve a wide variety of parameter estimation problems, especially when the likelihood function can be simplified by assuming the existence of additional but *missing* data $\mathcal{Y} = \{y_i | i = 1, \ldots, n\}$ corresponding to $\mathcal{X}$. The observations together with the missing data are called *complete data*. The EM algorithm maximizes the log-likelihood of the incomplete data by exploiting the relationship between the complete data and the incomplete data. In each iteration, two steps, called E-step and M-step, are involved. In the E-step, the EM algorithm determines the expectation of log-likelihood of the complete data based on the incomplete data and the current parameter

$$Q(\Theta|\Theta^{(t)}) = E\left(\log p(\mathcal{X}, \mathcal{Y}|\Theta) \middle| \mathcal{X}, \Theta^{(t)}\right) \tag{8.2}$$

In the M-step, the algorithm determines a new parameter maximizing Q

$$\Theta^{(t+1)} = \arg \max_{\Theta} Q(\Theta|\Theta^{(t)}) \tag{8.3}$$

Each iteration is guaranteed to increase the likelihood, and finally the algorithm converges to a local maximum of the likelihood function.

Clearly, the missing data $\mathcal{Y}$ has strong affects on the performance of the EM algorithm since the optimal parameter Θ^* is obtained through maximizing $E\left(\log p(\mathcal{X}, \mathcal{Y}|\Theta)\right)$. For example, the EM algorithm finds a *local* maximum of the likelihood function, which depends on the choice of $\mathcal{Y}$. Since the missing data $\mathcal{Y}$ is totally unknown and is "guessed" from the incomplete data, how can we choose a suitable $\mathcal{Y}$ to make the solution more reasonable? This question is not addressed in the EM algorithm because the likelihood function does not reflect any influence of the missing data. In order to address the issue, a simple and direct method is to regularize the likelihood function with a suitable functional of the distribution of the complete data.

In this chapter, we introduce a regularized EM (REM) algorithm to address the above issue. The basic idea is to regularize the likelihood function with the mutual information between the observations and the missing data or the conditional entropy of the missing data given the observations. The intuition behind is that we hope that the missing data have little uncertainty given the incomplete data because the EM algorithm implicitly assumes a strong relationship between the missing data and the incomplete data. When we apply the regularized EM algorithm to fit the finite mixture model, the new method can efficiently fit the models and effectively simplify over-complicated models.

162

The rest of the chapter is organized as follows. Section 8.2 introduces a regularized EM algorithm. In Section 8.3, we apply the regularized EM algorithm to the finite mixture model. Section 8.4 gives a demonstration of the new method on Gaussian mixtures. Section 8.5 concludes the chapter with some directions of future research.

8.2   The Regularized EM Algorithm

Simply put, the regularized EM algorithm tries to optimize the penalized likelihood

$$\tilde{L}(\Theta; \mathcal{X}) = L(\Theta; \mathcal{X}) + \gamma P(\mathcal{X}, \mathcal{Y}|\Theta) \qquad (8.4)$$

where the regularizer P is a functional of the distribution of the complete data given Θ and the positive value γ is the so-called regularization parameter that controls the compromise between the degree of regularization of the solution and the likelihood function.

As mentioned before, the EM algorithm assumes the existence of missing data. Intuitively, we would like to choose the missing data that has a strong (probabilistic) relation with the observations, which implies that the missing data has little uncertainty given the observations. In other words, the observations contain a lot of *information* about the missing data and we can infer the missing data from the observations with a small error rate. In general, the information about one object contained in another object can be measured by either Kolmogorov (or algorithmic) mutual information based on the theory of Kolmogorov complexity or Shannon mutual information based on Shannon information theory.

Both the theory of Kolmogorov complexity [Li and Vitányi, 1997] and Shannon information theory [Shannon, 1948] aim at providing a means for measuring the quantity of *information* in terms of *bit*. In the theory of Kolmogorov complexity, the *Kolmogorov complexity* (or *algorithmic entropy*) $K(x)$ of a finite binary string x is defined as the length of a shortest binary program p to compute x on an appropriate universal computer, such as a universal Turing machine. The conditional Kolmogorov complexity $K(x|y)$ of x relative to y is defined similarly as the length of a shortest program to compute x if y is furnished as an auxiliary input to the computation. The *Kolmogorov* (or *algorithmic*) *mutual information* is defined as $I(x:y) = K(y) - K(y|x, K(x))$ that is the information about x contained in y. Up to an additive constant term, $I(x:y) = I(y:x)$. Although $K(x)$ is the ultimate lower bound of any other complexity measures, $K(x)$ and related quantities are not Turing computable. Therefore, we can only try to approximate these quantities in practice.

In Shannon information theory, the quantity entropy plays a central role as measures of information, choice and uncertainty. Mathematically, Shannon's entropy of a discrete random variable X with a probability mass function $p(x)$ is defined as [Shannon, 1948]

$$H(X) = -\sum_x p(x) \log p(x) \qquad (8.5)$$

Entropy is the number of bits on the average required to describe a random variable. In fact, entropy is the minimum descriptive complexity of a random variable [Kolmogorov, 1965]. Consider two random variables X and Y with a joint distribution $p(x, y)$ and marginal distributions $p(x)$ and $p(y)$, respectively. The *conditional entropy* $H(X|Y)$ is defined as

$$H(X|Y) = \sum_y p(y)H(X|Y = y) = -\sum_x \sum_y p(x, y) \log p(x|y) \qquad (8.6)$$

which measures how uncertain we are of X on the average when we know Y. The *mutual information* $I(X;Y)$ between X and Y is the relative entropy (or Kullback-Leibler distance) between the joint distribution $p(x, y)$ and the product distribution $p(x)p(y)$

$$I(X;Y) = \sum_x \sum_y p(x, y) \log \frac{p(x, y)}{p(x)p(y)} \qquad (8.7)$$

which is symmetric. Note that when X and Y are independent, Y can tell us nothing about X and it is easy to show $I(X;Y) = 0$ in this case. Besides, the relationship between entropy and mutual information $I(X;Y) = H(X) - H(X|Y) = H(Y) - H(Y|X)$ demonstrates that the mutual information measures the amount of information that one random variable contains about another one. For continuous random variables, the summation operation is replaced with integration in the definitions of entropy and related notions.

Clearly, the theory of Kolmogorov complexity and Shannon information theory are fundamentally different although they share the same purpose. Shannon information theory considers the uncertainty of the population but ignores each individual. On the other hand, the theory of Kolmogorov complexity considers the complexity of a single object in the ultimate compressed version irrespective of the manner in which the object arose. Besides, Kolmogorov thinks that information theory must precede probability theory, and not be based on it [Kolmogorov, 1983b]. To regularize the likelihood function, we prefer Shannon mutual information to Kolmogorov mutual information because we cannot precede probability theory since the goal is just to estimate the parameters of distributions. Besides, we do consider the characteristics of the population of missing data rather than a single object in this case. Moreover, Kolmogorov complexity is not computable and we have to approximate it in applications. In fact, entropy is a popular approximation to Kolmogorov complexity in practice because it is a computable upper bound of Kolmogorov complexity [Kolmogorov, 1983a].

With Shannon mutual information as the regularizer, we have the regularized likelihood

$$\widetilde{L}(\Theta; \mathcal{X}) = L(\Theta; \mathcal{X}) + \gamma I(X; Y|\Theta) \tag{8.8}$$

where X is the random variable of observations and Y is the random variable of missing data. Because we usually do not know much about the missing data, we may naturally assume that Y follows a uniform distribution and thus $H(Y)$ is a constant value given the range of Y. Since $I(X; Y) = H(Y) - H(Y|X)$, we may also use the following regularized likelihood

$$\widetilde{L}(\Theta; \mathcal{X}) = L(\Theta; \mathcal{X}) - \gamma H(Y|X; \Theta) \tag{8.9}$$

Fano's inequality [Cover and Thomas, 1991] provides us another evidence that the conditional entropy $H(Y|X)$ could be a good regularizer here. Suppose we know a random variable X and we wish to guess the value of the correlated random variable Y that takes values in $\mathfrak{Y}$. Fano's inequality relates the probability of error in guessing the random variable Y to its conditional entropy $H(Y|X)$. Suppose we employ a function $\hat{Y} =$

$f(X)$ to estimate Y. Define the probability of error $P_e = \Pr\{\hat{Y} \neq Y\}$. Fano's inequality is

$$H(P_e) + P_e \log(|\mathfrak{Y}| - 1) \geq H(Y|X) \qquad (8.10)$$

This inequality can be weakened to

$$1 + P_e \log |\mathfrak{Y}| \geq H(Y|X) \qquad (8.11)$$

Note that $P_e = 0$ implies that $H(Y|X) = 0$. In fact, $H(Y|X) = 0$ if and only if Y is a function of X [Cover and Thomas, 1991]. Fano's inequality indicates that we can estimate Y with a low probability of error only if the conditional entropy $H(Y|X)$ is small. Thus, the conditional entropy of missing variable given the observed variable(s) is clearly a good regularizer for our purpose.

To optimize (8.8) or (8.9), we only need slightly modify the M-step of the EM algorithm. Instead of (8.3), we use

$$\Theta^{(t+1)} = \arg\max_{\Theta} \tilde{Q}(\Theta|\Theta^{(t)}) \qquad (8.12)$$

where

$$\tilde{Q}(\Theta|\Theta^{(t)}) = Q(\Theta|\Theta^{(t)}) + \gamma I(X;Y|\Theta) \qquad (8.13)$$

or

$$\tilde{Q}(\Theta|\Theta^{(t)}) = Q(\Theta|\Theta^{(t)}) - \gamma H(Y|X;\Theta) \qquad (8.14)$$

The modified algorithm is called the regularized EM (REM) algorithm. We can easily prove the convergence of the REM algorithm in the framework of proximal point algorithm [Bertsekas, 1999]. For the objective function $f(\Theta)$, a generalized proximal point algorithm is defined by the iteration

$$\Theta^{(t+1)} = \arg\max_{\Theta}\{f(\Theta) - \beta_t d(\Theta, \Theta^{(t)})\} \qquad (8.15)$$

where $d(\Theta, \Theta^{(t)})$ is a distance-like penalty function (i.e. $d(\Theta, \Theta^{(t)}) \geq 0$ and $d(\Theta, \Theta^{(t)}) = 0$ if and only if $\Theta = \Theta^{(t)}$), and β_t is a sequence of positive numbers. It is easy to show that the objective function $f(\Theta)$ increases with the iteration (8.15). In [Chretien and Hero, 2000], it was shown that EM is a special case of proximal point algorithm implemented with

$\beta_t = 1$ and a Kullback-type proximal penalty. In fact, the M-step of the EM algorithm can be represented as

$$\Theta^{(t+1)} = \arg\max_{\Theta} \left\{ L(\Theta; \mathcal{X}) - E\left[\log \frac{p(\mathcal{Y}|\mathcal{X}; \Theta^{(t)})}{p(\mathcal{Y}|\mathcal{X}; \Theta)} \middle| \mathcal{X}, \Theta^{(t)} \right] \right\} \quad (8.16)$$

Thus, we can immediately prove the convergence of the REM algorithm by replacing $L(\Theta; \mathcal{X})$ with $\tilde{L}(\Theta; \mathcal{X})$ in (8.16). Because the regularization term in (8.8) (or (8.9)) biases the searching space to some extent, we expect that the REM algorithm also converges faster than the plain EM algorithm, which will be confirmed in the experiments.

8.3   Finite Mixture Model

In this section, we apply the regularized EM algorithm to fit the finite mixture model. The finite mixture model arises as the fundamental model naturally in the areas of statistical machine learning. With the finite mixture model, we assume that the density associated with a population is a finite mixture of densities. Finite mixture densities can naturally be interpreted as that we have m component densities mixed together with mixing coefficients $\alpha_k, k = 1, \ldots, m$, which can be thought of as the *a priori* probabilities of each mixture component c_k, i.e. $\alpha_k = p(c_k)$. The mixture probability density functions have the form

$$p(x|\Theta) = \sum_{k=1}^{m} \alpha_k p(x|\theta_k) \quad (8.17)$$

where the parameters are $\Theta = (\alpha_1, \ldots, \alpha_m, \theta_1, \ldots, \theta_m)$ such that $\alpha_k \geq 0, k = 1, \ldots, m$ and $\sum_{k=1}^{m} \alpha_k = 1$; and each p is the density function of the component c_k that is parameterized by θ_k. [1]

For the finite mixture model, we usually employ the category information $\mathcal{C}$ associated with the observations $\mathcal{X}$ as the missing data, which

---

[1] Here, we assume that all components have the same form of density for simplicity. More generally, the densities do not necessarily need belong to the same parametric family.

indicates which component in the mixture produces the observation. In this section, we use the conditional entropy as the regularizer in particular. The reason will be clear later. Let C be a random variable taking values in $\{c_1, c_2, \ldots, c_m\}$ with probabilities $\alpha_1, \alpha_2, \ldots, \alpha_m$. Thus, we have

$$
\widetilde{L}(\Theta; \mathcal{X}) = L(\Theta; \mathcal{X}) - \gamma H(C|X; \Theta)
$$

$$
= \sum_{i=1}^{n} \log \sum_{k=1}^{m} \alpha_k p(x_i|\theta_k)
$$

$$
+ \gamma \int \sum_{k=1}^{m} \frac{\alpha_k p(x|\theta_k)}{p(x|\Theta)} \log \left(\frac{\alpha_k p(x|\theta_k)}{p(x|\Theta)} \right) p(x|\Theta) dx
$$

The corresponding $\widetilde{Q}$ is

$$
\widetilde{Q}(\Theta|\Theta^{(t)}) = \sum_{k=1}^{m} \sum_{i=1}^{n} \log(\alpha_k) p(c_k|x_i; \Theta^{(t)})
$$

$$
+ \sum_{k=1}^{m} \sum_{i=1}^{n} \log(p(x_i|\theta_k)) p(c_k|x_i; \Theta^{(t)})
$$

$$
+ \gamma \int \sum_{k=1}^{m} \frac{\alpha_k p(x|\theta_k)}{p(x|\Theta)} \log \frac{\alpha_k p(x|\theta_k)}{p(x|\Theta)} p(x|\Theta) dx
$$

In order to find $\alpha_k, k = 1, \ldots, m$, we introduce a Lagrangian

$$
\mathcal{L} = \widetilde{Q}(\Theta|\Theta^{(t)}) - \lambda \left(\sum_{k=1}^{c} \alpha_k - 1 \right) \tag{8.18}
$$

with multiplier λ for the constraint $\sum_{k=1}^{m} \alpha_k = 1$. Solving the Lagrangian $\mathcal{L}$, we obtain (for details, see Appendix B.1)

$$
\alpha_k^{(t+1)}
$$

$$
= \frac{\sum_{i=1}^{n} p(c_k|x_i; \Theta^{(t)}) + \gamma \int p(c_k|x; \Theta) \log p(c_k|x; \Theta) p(x|\Theta) dx}{\sum_{k=1}^{m} \sum_{i=1}^{n} p(c_k|x_i; \Theta^{(t)}) + \gamma \sum_{k=1}^{m} \int p(c_k|x; \Theta) \log p(c_k|x; \Theta) p(x|\Theta) dx} \tag{8.19}
$$

For simplicity, we employ $\frac{1}{n} \sum_{i=1}^{n} p(c_k|x_i; \Theta^{(t)}) \log p(c_k|x_i; \Theta^{(t)})$ to approximate the integrals in the above equation. Therefore, we obtain

$$
\alpha_k^{(t+1)} = \frac{\displaystyle\sum_{i=1}^{n} p(c_k|x_i; \Theta^{(t)})(1 + \gamma \log p(c_k|x_i; \Theta^{(t)}))}{\displaystyle\sum_{i=1}^{n} \sum_{k=1}^{m} p(c_k|x_i; \Theta^{(t)})(1 + \gamma \log p(c_k|x_i; \Theta^{(t)}))} \tag{8.20}
$$

where we drop the coefficient $\frac{1}{n}$ associated with $\log p(c_k|x_i; \Theta^{(t)})$ since we can adjust the regularization parameter γ to match it.

To find $\theta_k, k = 1, \ldots, m$, we take the derivatives of $\widetilde{Q}$ with respect to θ_k

$$\frac{\partial \widetilde{Q}(\Theta|\Theta^{(t)})}{\partial \theta_k} = 0 \qquad k = 1, \ldots, m$$

For exponential families, it is possible to get an analytical expression for θ_k, as a function of everything else. Suppose that $p(x|\theta_k)$ has the regular exponential-family form [Barndorff-Nielsen, 1978]:

$$p(x|\theta_k) = \varphi^{-1}(\theta_k)\psi(x)e^{\theta_k^T t(x)} \tag{8.21}$$

where θ_k denotes an $r \times 1$ vector parameter, $t(x)$ denotes an $r \times 1$ vector of sufficient statistics, the superscript T denotes matrix transpose, and $\varphi(\theta_k)$ is given by

$$\varphi(\theta_k) = \int \psi(x)e^{\theta_k^T t(x)} dx \tag{8.22}$$

The term "regular" means that θ_k is restricted only to a convex set Ω such that equation (8.21) defines a density for all θ_k in Ω. Such parameters are often called natural parameters. The parameter θ_k is also unique up to an arbitrary non-singular $r \times r$ linear transformation, as is the corresponding choice of $t(x)$. For example, *expectation parametrization* employs $\phi(\theta_k) = E(t(x)|\theta_k)$, which is a both-way continuously differentiable mapping [Barndorff-Nielsen, 1978].

For exponential families, we have (see Appendix B.2 for details)

$$\phi(\theta_\ell) = \frac{\sum_{i=1}^n t(x_i)p(c_\ell|x_i; \Theta^{(t)}) + \gamma \int t(x)p(c_\ell|x; \Theta) \log p(c_\ell|x; \Theta)p(x|\Theta)dx}{\sum_{i=1}^n p(c_\ell|x_i; \Theta^{(t)}) + \gamma \int p(c_\ell|x; \Theta) \log p(c_\ell|x; \Theta)p(x|\Theta)dx} \tag{8.23}$$

Similar to (8.20), we obtain

$$\phi^{(t+1)}(\theta_k) = \frac{\displaystyle\sum_{i=1}^n t(x_i)p(c_k|x_i; \Theta^{(t)})(1 + \gamma \log p(c_k|x_i; \Theta^{(t)}))}{\displaystyle\sum_{i=1}^n p(c_k|x_i; \Theta^{(t)})(1 + \gamma \log p(c_k|x_i; \Theta^{(t)}))} \tag{8.24}$$

169

In particular, we have

$$\mu_k^{(t+1)} = \frac{\sum_{i=1}^{n} x_i p(c_k|x_i; \Theta^{(t)})(1 + \gamma \log p(c_k|x_i; \Theta^{(t)}))}{\sum_{i=1}^{n} p(c_k|x_i; \Theta^{(t)})(1 + \gamma \log p(c_k|x_i; \Theta^{(t)}))} \qquad (8.25)$$

$$\Sigma_k^{(t+1)} = \frac{\sum_{i=1}^{n} (x_i - \mu_k)(x_i - \mu_k)^T p(c_k|x_i; \Theta^{(t)})(1 + \gamma \log p(c_k|x_i; \Theta^{(t)}))}{\sum_{i=1}^{n} p(c_k|x_i; \Theta^{(t)})(1 + \gamma \log p(c_k|x_i; \Theta^{(t)}))}$$

$$(8.26)$$

for a Gaussian mixture $p(x) = \sum_{k=1}^{m} \alpha_k N(\mu_k, \Sigma_k)$.

When we apply the EM algorithm to fit the finite mixture model, we have to determine the number of components, which is usually referred as to model selection. Because the maximized likelihood is a non-decreasing function of the number of components [Figueiredo and Jain, 2002], the plain EM algorithm cannot reduce a specified over-complicated model to a simpler model by itself. That is, if a larger number of components is specified, the plain EM algorithm cannot reduce it to the true but smaller number of components (i.e. a simpler model). Because over-complicated models introduce more uncertainty, [2] we expect that the REM algorithm in contrast will be able to automatically simplify over-complicated models to simpler ones through reducing the uncertainty of missing data. That is, excess components will be purged out of the mixture model by assigning near-zero probabilities. Besides, note the conditional entropy of category information C given X

$$H(C|X) = -\int \sum_{k=1}^{m} p(c_k|x) \log(p(c_k|x)) p(x) dx \qquad (8.27)$$

is a non-decreasing function of the number of components because a larger m implies more choices and a larger entropy [Shannon, 1948]. In fact, $H(C|X)$ is minimized to 0 if $m = 1$, i.e. all data are from the same component. Thus, the term $-\gamma H(C|X)$ in $\widetilde{L}(\Theta; \mathcal{X})$ would support the

---

[2] The more choices, the more entropy [Shannon, 1948].

merge of the components to reduce the entropy in the iterations of REM. On the other hand, the term $L(\Theta; \mathcal{X})$ supports keeping the number of components as large as possible to achieve a high likelihood. Finally, the REM algorithm reaches a balance between the likelihood and the conditional entropy and it reduces the number of components to some extent.

The model selection problem is an old problem and many criteria/methods have been proposed, such as Akaike's information criterion (AIC) [Akaike, 1973], Bayesian inference criterion (BIC) [Schwarz, 1978], Cheeseman-Stutz criterion [Cheeseman and Stutz, 1995], minimum message length (MML) [Wallace and Boulton, 1968], and minimum description length (MDL) [Rissanen, 1985]. However, we do not attempt to compare our method with the aforementioned methods because the goal of our method is to reduce the uncertainty of missing data rather than to determine the number of components. In fact, simplifying an over-complicated model is only a byproduct of our method obtained through reducing the uncertainty of missing data. Besides, our method is not a comprehensive method to determine the number of components since it cannot extend an over-simple model to the true model.

8.4   Demonstration

In this section, we present an example to illustrate the performance of the REM algorithm on a two-dimensional Gaussian mixture. The mixture contains six components, each of which has 300 samples. The data is shown in Figure 8.1. In the experiments, we use k-means to give the initial partition. The stop criterion in iterations is that the increase in the regularized log-likelihood (8.9) is less than 10^{-7}. In the experiments, we test the REM algorithm with different numbers of components and regularization factor γ. Note that the REM algorithm reduces to the plain EM algorithm when γ is set to 0. With each setting, we run the algorithm 30 times. The medians of the results are reported here.

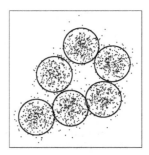

Figure 8.1: The simulated Gaussian mixture.

To measure the quality of learned models, we employ BIC/MDL [3] [Schwarz, 1978, Rissanen, 1985] here for simplicity. Let v be the number of independent parameters to be estimated in the model. [4] BIC can be approximated by

$$BIC \approx L(\hat{\Theta}) - \frac{1}{2}v \log n \qquad (8.28)$$

A large BIC score indicates that the model has a large posteriori and thus is most likely close to the true model. As shown in Figure 8.2, the REM algorithm achieves much larger BIC scores than the plain EM algorithm (i.e. the $\gamma = 0$ case) when the number of components is incorrectly specified. When the specified number of components is correct (i.e. $m = 6$), the plain EM and REM obtain similar BIC scores. We also observe that, if a suitable γ is employed, the REM algorithm may achieve a higher BIC score than the EM algorithm even when the number of components is correctly set for the EM algorithm. For example, the REM algorithm achieves a higher BIC score with $\gamma = 0.05$ and $m = 8$ than that of the EM algorithm with $m = 6$. This study also suggests that we may choose γ by BIC/MDL. Further research on determining the optimal γ is in progress.

Besides BIC/MDL scores, we also investigate the number of components in the learned models. In this study, we regard a component as purged out of the model if its priori probability is less than 0.01. The

---

[3]BIC coincides with the two-stage form of MDL [Hansen and Yu, 2001].

[4]We consider only the parameters of the components with non-zero probabilities.

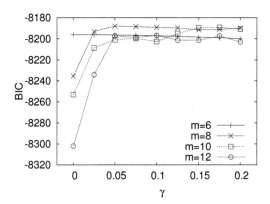

Figure 8.2: The BIC scores of the learned models. Here, m is the specified number of components and γ is the regularization factor.

(median of) learned numbers of components are shown in Figure 8.3. As shown in the figure, the REM algorithm can usually reduce an incorrectly specified number of components to the correct one (i.e. 6). We also observe that the REM algorithm does not reduce the models to over-simplified ones (e.g. the learned number of components is less than 6) in all cases. It is well-known that the plain EM algorithm may also return empty clusters (corresponding to components with zero probability), which is confirmed in our experiments. For $m = 10$ and $m = 12$, we observe that the EM algorithm may return fewer (say 9 or 11) components. Compared with the true model, however, it is still far from perfection.

It is known that the EM algorithm may converge very slowly in practice. In the experiments, we find that the REM algorithm converges much faster than the EM algorithm as shown in Figure 8.4. The reason may be that the regularization biases the search space toward more likely regions so that it improves the efficiency of iterations. Interestingly, the number of iterations seems to decrease with the increase of γ.

Finally, we give a graphical representation of iterations of the REM

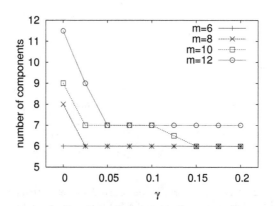

Figure 8.3: Number of components.

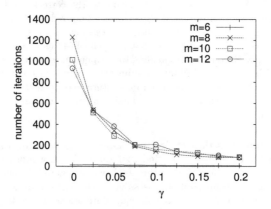

Figure 8.4: Number of iterations.

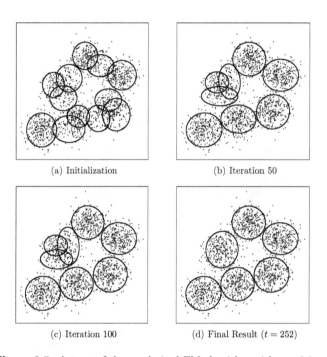

(a) Initialization

(b) Iteration 50

(c) Iteration 100

(d) Final Result ($t = 252$)

Figure 8.5: A trace of the regularized EM algorithm with $\gamma = 0.1$ and $m = 12$.

algorithm in Figure 8.5. Here, we set $\gamma = 0.1$ and $m = 12$. After 50 iterations, the estimated model can already describe the shape of the data well. Finally, the REM algorithm converges at iteration 252 with six components that are very close to the true model. The extra components (not represented in the figure) are successively purged from the model due to their zero *a priori* probabilities.

8.5 Conclusion

We have proposed a regularized EM algorithm to control the uncertainty of missing data. The REM algorithm tries to maximize the likelihood and the information about the missing data contained in the observations. Besides reducing the uncertainty of missing data, the proposed method maintains the advantage of the conventional EM algorithm. When we apply the regularized EM algorithm to fit the finite mixture model, it can efficiently fit the models and effectively simplify over-complicated models. The convergence properties of the REM algorithm would be an interesting future research topic.

Chapter 9

Conclusions

Computational methods have become a central feature of research and discovery in the life sciences. With the fast development of molecular biology, there are various challenging computational problems waiting for researchers. In this book, we try to provide solutions/tools for several important problems with the help of machine learning techniques. In particular, we present the edit kernels for use with SVMs to accurately recognize translation initiation sites in eukaryotic mRNAs. For cancer classification with gene expression profiling, we propose GLDA to overcome the curse of dimensionality and the small sample size problem. For gene expression analysis, we develop both clustering and biclustering methods. More precisely, we present minimum entropy clustering method that does not need the number of clusters in advance, is able to find clusters with arbitrary shape, and can effectively detect outliers. We also propose a novel approach for biclustering that in principle can discover all computable patterns in gene expression data based on the theory of Kolmogorov complexity. Beyond these single gene expression dataset analysis methods, we also develop an integrative approach for analyzing multiple microarray gene expression datasets together. Besides DNA sequence and gene expression data, we also work with restriction map data. Especially, we develop a quantile method for sizing optical maps. Due to the wide applications of the EM algorithm in bioinformatics, we also develop the regularized EM algorithm to control the uncertainty of missing data.

In these work, both supervised (e.g. SVM and LDA) and unsupervised (e.g. clustering and biclustering) learning methods are employed. Due to the multidisciplinary nature of machine learning, various techniques from different disciplines are used in the development of our methods, for example mathematics, statistics, information theory, computer science, etc. Obviously, the choice of specific technique depends on the background and properties of the molecular biology problems. In fact, in order to successfully apply machine learning methods to biology problems, it is also very important to understand the biological background and incorporate important priori information into the learning models.

Through these work, it is clear that machine learning is a useful computational tool for biology research. Moreover, I believe that machine learning will be an integrated part of the research and discovery of biology and will greatly speedup our research.

So far, our work on computational biology with machine learning mainly focuses on data analysis. In the future, an interesting research topic is how to employ machine learning to more directly help us understand the biological procedures. For example, we have build an accurate model to recognize TISs in eukaryotic mRNAs. Although such a model is already useful in terms of predicting unknown TISs in new sequences, it is more interesting and useful if we can discover the translation initiation mechanisms of the ribosome based on this model. To point out a potential future research direction, let us recall the old story of the drunk searching for his key:

A man watched a drunk searching for something on the ground around a street lamp.

"What have you lost?" he asked.

"My key," said the drunk.

"Where exactly did you drop it?"

"Up the road a bit," the drunk replied.

"Then why aren't you looking for it there?"

"Cause there's more light here than up the road."

The drunk seems to have acted weirdly. Ironically, we might be doing research in the same way as the drunk. Our research goal is to understand the unknown ground truth of biological systems, which is in the "dark". Just as the drunk looks for his key where there is the most light, researchers are drawn to their "street lamp" — experimental data and previously published work. Starting from the current knowledge (the street lamp), we wish understand truth more and more (i.e., get more light). Finally, the key is in the dim halo of of the street lamp and we may find it.

Unfortunately, this procedure could take very long time. There are many paths starting from the street lamp. Some short paths lead us to the key quickly. Some paths may be much longer. Some paths even lead us to wrong directions. Is there some approach that could help us to follow the short paths more likely?

Suppose that biological research is a process that discovers or understands biological procedures from the experimental data. Mathematically, we may naturally assume that the most possible biological procedure producing an experimental phenomenon has the largest *a posteriori* probability:

$$P(procedure|phenomenon) = \frac{P(procedure)\,P(phenomenon|procedure)}{P(phenomenon)}$$

Given a biological phenomenon, therefore, we should investigate the procedures that easily occur (i.e. with large $P(procedure)$) and likely produce the phenomenon (i.e. with large $P(phenomenon|procedure)$). With the above probabilistic representation, we may see that statistical machine learning may help us to find the key quickly.

In practice, a big challenge is how to propose/discover the most likely procedure from experimental data for biologists to do further investigation or verification. With machine learning, we may first try to fit a model to the experimental data. If the model achieves a high accuracy, we could assume that the model captures some characteristics of the underlying biological procedure. Then, we may try to extract some useful information from the learned model. However, the model is usually a black box. For example, we build a highly accurate support vector machine to predict translation initiation sites. But we do not exactly know why it works well and what it learns from the data. To extract useful information, we have to "decode" the model, which would be an interesting research topic. In the case of predicting TISs, the learned support vectors could be a good clue to discover the working scheme of the ribosome. By aligning the positive support vectors, for example, we may find some conserved sites that attract the 60S subunit of the ribosome to initiate the translation of the downstream. On the other hand, we can perhaps also align negative support vectors to find the conserved sites that push 60S subunit away so that the translation cannot be initiated.

Note that it is very likely that we have to repeat the above process many times during the investigation. First, some experiments are designed to verify a biology hypotheses and then some models are fitted on the experimental results. By analyzing the learned models, we may hopefully discover some new biological information. With the new knowledge, we will be stimulated to generate new hypotheses and develop experiments to test them. The resulting experimental data will, in turn, be used to generate more refined models that will improve overall understanding. This procedure is repeated until we fully understand the target biology process.

Finally, we have to be aware that this research approach may not work sometime since statistical significance does not always imply biological significance. However, we expect that the approach works reasonably well and is worth trying.

Appendix A

Matrix Algebra

A.1   Notation

Unless explicitly stated, all vectors are column vectors and belong to $\mathbb{R}^n$. For any $x \in \mathbb{R}^n$, x^T denotes the transpose of x, which is an n-dimensional row vector. The ith coordinate/component of x is denoted by x_i. The notations $x > 0$ and $x \geq 0$ indicate that all coordinates of x are positive and nonnegative, respectively. For any two vectors x and y, the notation $x > y$ means that $x - y > 0$. The notations $x \geq y$, $x < y$, and $x \leq y$ are to be interpreted accordingly. The vectors $x_1, \ldots, x_m \in \mathbb{R}^n$ are called *linearly independent* if there exists no set of scalars $\alpha_1, \ldots, \alpha_m$ such that $\sum_{k=1}^{m} \alpha_k x_k = 0$, unless $\alpha_k = 0$ for each k. The *inner product* of two vectors $x, y \in \mathbb{R}^n$ is defined by $x^T y = \sum_{i=1}^{n} x_i y_i$. Any two vectors $x, y \in \mathbb{R}^n$ satisfying $x^T y = 0$ are called *orthogonal*. If x, y are orthogonal and have length one, they are called *orthonormal*.

For any matrix A, we use A_{ij} to denote its ijth element. The *transpose* of A, denoted by A^T, is defined by $A_{ij}^T = A_{ji}$. Let A be an $m \times n$ matrix. The *range space* of A is the set of all vectors $y \in \mathbb{R}^m$ such that $y = Ax$ for some $x \in \mathbb{R}^n$. The *null space* or *kernel* of A is the set of all vectors $x \in \mathbb{R}^n$ such that $Ax = 0$. The *rank* of A is the dimensionality of the range space of A. The rank of A is equal to the maximal number of linearly independent columns (or rows) of A. We say that A has *full rank* if its rank is equal to $\min\{m, n\}$. Let A be an $n \times n$ square matrix. A square matrix A is *symmetric* if $A^T = A$. If $A_{ij} = 0$ whenever $i \neq j$, we say that A is *diagonal*. The identity matrix is denoted by I. The trace of a square matrix A, denoted by $tr(A)$, is the sum of A's diagonal elements. The *determinant* of A is denoted by $\det(A)$ or $|A|$. A is called *singular* if its determinant is zero. Otherwise it is called nonsignular or invertible. A square matrix is nonsignular if and only if it is of full rank. Assuming that A is nonsignular, the matrix B such that $AB = BA = I$ is called the inverse of A. The inverse of a nonsignular matrix A is unique and denoted by A^{-1}. A square matrix A is called idempotent if it satisfies $AA = A$. An idempotent matrix A is called an orthogonal projector if $A = A^T$. A symmetric $n \times n$ matrix A is called *positive definite* if $x^T A x > 0$ for all $x \in \mathbb{R}^n$, $x \neq 0$. It is called *positive semidefinite* if $x^T A x \geq 0$ for all $x \in \mathbb{R}^n$. The notations $A \succ 0$ and $A \succeq 0$ means that A is positive definite and positive semidefinite, respectively.

A.2 Eigenvalues and Eigenvectors

The *characteristic polynomial* ϕ of an $n \times n$ matrix A is defined as $\phi(\lambda) = |\lambda I - A|$. The n (possibly repeated or complex) roots of ϕ are called the *eigenvalues* (also called *characteristic roots*) of A. Corresponding to each eigenvalue λ, the nonzero vectors x (with possibly complex coordinates) satisfying $Ax = \lambda x$ are called the *eigenvectors* (also called *characteristic vectors*) of A associated with λ.

Proposition A.1 *Let A be an $n \times n$ matrix.*

1. *A is singular if and only if it has an eigenvalue that is equal to zero.*

2. *If A is nonsingular, then the eigenvalues of A^{-1} are the reciprocals of the eigenvalues of A.*

3. *If T is nonsingular matrix, then the eigenvalues of A and TAT^{-1} coincide.*

4. *The eigenvalues of A and A^T coincide.*

5. *The eigenvalues of A^k are equal to $\lambda_1^k, \ldots, \lambda_n^k$, where $\lambda_1, \ldots, \lambda_n$ are the eigenvalues of A.*

6. *For any scalar c, the eigenvalues of $cI + A$ are equal to $c + \lambda_1, \ldots, c + \lambda_n$, where $\lambda_1, \ldots, \lambda_n$ are the eigenvalues of A.*

Symmetric matrices have some special properties in regard to eigenvalues and eigenvectors. Furthermore, the widespread use of symmetric matrices in statistics make it worthwhile to discuss these properties in detail.

Proposition A.2 *Let A be a symmetric $n \times n$ matrix.*

1. *The eigenvalues of A are real.*

2. *The matrix A has a set of n mutually orthogonal, real, and nonzero eigenvectors.*

3. *The matrix A is positive semidefinite (respectively, positive definite) if and only if all of its eigenvalues are nonnegative (respectively, positive).*

4. *The rank of A is equal to the number of nonzero eigenvalues.*

5. *Suppose that $x_1, \ldots, x_n$ are the orthonormal eigenvectors of A. Then*

$$A = \sum_{i=1}^{n} \lambda_i x_i x_i^T \tag{A.1}$$

where λ_i is the eigenvalue corresponding to x_i.

A.3   Matrix Derivatives

The problem of differentiating a function of a matrix $f(A)$ with respect to the elements of the matrix A is an important one in multivariate analysis. We collect here several important formulae of differentiation. Let $A = A(x)$ be a nonsingular matrix such that its elements depend on a scalar x. Then

$$\frac{\partial A^{-1}}{\partial x} = -A^{-1}\frac{\partial A}{\partial x}A^{-1} \tag{A.2}$$

Let $\partial/\partial A$ be a matrix of derivative operators. For trace functions, we have

$$\frac{\partial}{\partial A}tr(AB) = B^T \tag{A.3}$$

$$\frac{\partial}{\partial A}tr(A^T B) = B \tag{A.4}$$

$$\frac{\partial}{\partial A}tr(A^T BA) = (B + B^T)A \tag{A.5}$$

$$\frac{\partial}{\partial A^T}tr(A^T BA) = A^T(B + B^T) \tag{A.6}$$

$$\frac{\partial}{\partial A}tr((A^T BA)^{-1}(A^T CA)) = -2BA(A^T BA)^{-1}(A^T CA)(A^T BA)^{-1}$$
$$+ 2CA(A^T BA)^{-1} \tag{A.7}$$

For determinant functions, we have

$$\frac{\partial}{\partial A}|A| = |A|(A^{-1})^T \tag{A.8}$$

$$\frac{\partial}{\partial A}\ln|A| = (A^{-1})^T \tag{A.9}$$

$$\frac{\partial}{\partial A}|A^T BA| = 2|A^T BA|BA(A^T BA)^{-1} \tag{A.10}$$

$$\frac{\partial}{\partial A}\ln|A^T BA| = 2BA(A^T BA)^{-1} \tag{A.11}$$

A.4 Generalized Inverse of Matrices

Definition A.3 (Generalized Inverse) *Let A be an $m \times n$ matrix. Then an $n \times m$ matrix A^- is said to be a* generalized inverse *(or* pseudo inverse*) of A if*

$$AA^- A = A \tag{A.12}$$

holds. A generalized inverse always exists although it is not unique in general.

If

$$A = \begin{pmatrix} X & Y \\ Z & W \end{pmatrix}$$

with X nonsingular and $rank(X) = rank(A)$, we have

$$A^- = \begin{pmatrix} X^{-1} & 0 \\ 0 & 0 \end{pmatrix}$$

as a special generalized inverse.

Proposition A.4 *Let A^- be any generalized inverse of the matrix A.*

1. *If A is nonsingular, $A^- = A^{-1}$ uniquely.*

2. *A^-A and AA^- are idempotent.*

3. *$rank(A) = rank(AA^-) = rank(A^-A)$*

4. *$rank(A) \leq rank(A^-)$*

5. *$A(A^TA)^-A^T$ is unique, i.e. invariant to the choice of $(A^TA)^-$.*

Definition A.5 (Moore-Penrose Inverse) *A matrix A^+ satisfying the following conditions is unique and is called the Moore-Penrose inverse of A:*

$$AA^+A = A \tag{A.13}$$
$$A^+AA^+ = A^+ \tag{A.14}$$
$$(A^+A)^T = A^+A \tag{A.15}$$
$$(AA^+)^T = AA^+ \tag{A.16}$$

Proposition A.6 *Let A be an $m \times n$ matrix.*

1. $(A^+)^+ = A$

2. $(A^+)^T = (A^T)^+$

3. $A^+ = (A^TA)^+A^T = A^T(AA^T)^+$

4. $(A^TA)^+ = A^+(A^T)^+$ and $(AA^T)^+ = (A^T)^+A^+$

5. If A is nonsingular, $A^+ = A^{-1}$.

6. If A is an orthogonal projector, $A^+ = A$.

7. $rank(A) = rank(AA^+) = rank(AA^+) = rank(A^+A)$

8. If $rank(A) = m$, $A^+ = A^T(AA^T)^{-1}$ and $AA^+ = I_m$.

9. If $rank(A) = n$, $A^+ = (A^TA)^{-1}A^T$ and $A^+A = I_n$.

10. If P and Q are orthogonal, $(PAQ)^+ = Q^{-1}A^+P^{-1}$.

Proposition A.7 *Let $X \neq 0$ be an $m \times n$ matrix and A be an $n \times n$ matrix. If $X^TXAX^TX = X^TX$, then $XAX^TX = X$ and $X^TXAX^T = X^T$*

Corollary A.8 *Let $X \neq 0$ be an $m \times n$ matrix and A be an $n \times n$ matrix. Then $AX^TX = BX^TX$ if and only if $AX^T = BX^T$*

Appendix B

Derivations of Some Equations

B.1 Derivation of Equation (8.19)

Consider the Lagrangian $\mathcal{L}$ in Equation (8.18), which has to be maximized with respect to α_ℓ and λ. At the stationary point, the derivatives of $\mathcal{L}$ with respect to $\alpha_\ell, \ell = 1, \ldots, m$, must vanish

$$
\frac{\partial \mathcal{L}}{\partial \alpha_\ell} = \frac{1}{\alpha_\ell} \sum_{i=1}^{n} p(c_\ell|x_i; \Theta^{(t)})
$$

$$
+ \gamma \int \left(p(x|\theta_\ell) \log \frac{\alpha_\ell p(x|\theta_\ell)}{p(x|\Theta)} + \frac{p(x|\theta_\ell)p(x|\Theta) - \alpha_\ell p^2(x|\theta_\ell)}{p^2(x|\Theta)} p(x|\Theta) \right) dx
$$

$$
+ \gamma \int \sum_{\substack{k=1 \\ k \neq \ell}}^{m} -\frac{\alpha_k p(x|\theta_k)p(x|\theta_\ell)}{p^2(x|\Theta)} p(x|\Theta) dx - \lambda = 0
$$

Note that

$$\int \left(\frac{p(x|\theta_\ell)p(x|\Theta) - \alpha_\ell p^2(x|\theta_\ell)}{p^2(x|\Theta)} + \sum_{\substack{k=1 \\ k \neq \ell}}^{m} -\frac{\alpha_k p(x|\theta_k)p(x|\theta_\ell)}{p^2(x|\Theta)} \right) p(x|\Theta)dx$$

$$= \int \left(p(x|\theta_\ell) - \sum_{k=1}^{m} \frac{\alpha_k p(x|\theta_k)p(x|\theta_\ell)}{p(x|\Theta)} \right) dx$$

$$= \int \left(p(x|\theta_\ell) - p(x|\theta_\ell)\frac{\sum_{k=1}^{m} \alpha_k p(x|\theta_k)}{p(x|\Theta)} \right) dx$$

$$= 0$$

So, we have

$$\frac{\partial \mathcal{L}}{\partial \alpha_\ell} = \frac{1}{\alpha_\ell}\sum_{i=1}^{n} p(c_\ell|x_i; \Theta^{(t)}) + \gamma \int p(x|\theta_\ell)\log\frac{\alpha_\ell p(x|\theta_\ell)}{p(x|\Theta)}dx - \lambda = 0$$

i.e.,

$$\sum_{i=1}^{n} p(c_\ell|x_i; \Theta^{(t)}) + \gamma \int \alpha_\ell p(x|\theta_\ell)\log p(c_\ell|x; \Theta)dx = \lambda\alpha_\ell$$

$$\sum_{i=1}^{n} p(c_\ell|x_i; \Theta^{(t)}) + \gamma \int p(c_\ell|x; \Theta)\log p(c_\ell|x; \Theta)p(x|\Theta)dx = \lambda\alpha_\ell$$

Summing both sides over ℓ, we get

$$\lambda = \sum_{\ell=1}^{m}\sum_{i=1}^{n} p(c_\ell|x_i; \Theta^{(t)}) + \gamma \int \sum_{\ell=1}^{m} p(c_\ell|x; \Theta)\log p(c_\ell|x; \Theta)p(x|\Theta)dx$$

Thus,

$$\alpha_\ell = \frac{\sum_{i=1}^{n} p(c_\ell|x_i; \Theta^{(t)}) + \gamma \int p(c_\ell|x; \Theta)\log p(c_\ell|x; \Theta)p(x|\Theta)dx}{\sum_{\ell=1}^{m}\sum_{i=1}^{n} p(c_\ell|x_i; \Theta^{(t)}) + \gamma \int \sum_{\ell=1}^{m} p(c_\ell|x; \Theta)\log p(c_\ell|x; \Theta)p(x|\Theta)dx}$$

B.2   Derivation of Equation (8.23)

First, we have

$$
\begin{aligned}
\frac{\partial p(x|\theta_\ell)}{\partial \theta_\ell} &= \left(\frac{\partial \theta_\ell^T t(x)}{\partial \theta_\ell} - \frac{\partial \varphi(\theta_\ell)}{\partial \theta_\ell} \varphi^{-1}(\theta_\ell) \right) \varphi^{-1}(\theta_\ell) \psi(x) e^{\theta_\ell^T t(x)} \\
&= \left(t(x) - \varphi^{-1}(\theta_\ell) \int \psi(x) t(x) e^{\theta_\ell^T t(x)} dx \right) p(x|\theta_\ell) \\
&= \left(t(x) - \int t(x) \left(\varphi^{-1}(\theta_\ell) \psi(x) e^{\theta_\ell^T t(x)} \right) dx \right) p(x|\theta_\ell) \\
&= \left(t(x) - E(t(x)|\theta_\ell) \right) p(x|\theta_\ell) \\
&= \left(t(x) - \phi(\theta_\ell) \right) p(x|\theta_\ell)
\end{aligned}
$$

where $\phi(\theta_\ell) = E(t(x)|\theta_\ell)$. With it, we obtain

$$
\frac{\partial H(C|X;\Theta)}{\partial \theta_\ell}
$$

$$
= -\int \left(\alpha_\ell \frac{\partial p(x|\theta_\ell)}{\partial \theta_\ell} \log \frac{\alpha_\ell p(x|\theta_\ell)}{p(x|\Theta)} + \frac{\alpha_\ell p(x|\Theta) \frac{\partial p(x|\theta_\ell)}{\partial \theta_\ell} - \alpha_\ell p(x|\theta_\ell) \frac{\partial p(x|\Theta)}{\partial \theta_\ell}}{p^2(x|\Theta)} p(x|\Theta) \right) dx
$$

$$
- \int \sum_{\substack{k=1 \\ k \neq \ell}}^{m} -\frac{\alpha_k p(x|\theta_k) \frac{\partial p(x|\Theta)}{\partial \theta_\ell}}{p^2(x|\Theta)} p(x|\Theta) dx
$$

$$
= -\int \left(\alpha_\ell \frac{\partial p(x|\theta_\ell)}{\partial \theta_\ell} \log \frac{\alpha_\ell p(x|\theta_\ell)}{p(x|\Theta)} + \alpha_\ell \frac{\partial p(x|\theta_\ell)}{\partial \theta_\ell} + \int \sum_{k=1}^{m} -\frac{\alpha_k p(x|\theta_k) \frac{\partial p(x|\Theta)}{\partial \theta_\ell}}{p(x|\Theta)} \right) dx
$$

$$
= -\int \left(\alpha_\ell \frac{\partial p(x|\theta_\ell)}{\partial \theta_\ell} \log \frac{\alpha_\ell p(x|\theta_\ell)}{p(x|\Theta)} + \alpha_\ell \frac{\partial p(x|\theta_\ell)}{\partial \theta_\ell} - \frac{\partial p(x|\Theta)}{\partial \theta_\ell} \int \sum_{k=1}^{m} \frac{\alpha_k p(x|\theta_k)}{p(x|\Theta)} \right) dx
$$

$$
= -\int \left(\alpha_\ell \frac{\partial p(x|\theta_\ell)}{\partial \theta_\ell} \log \frac{\alpha_\ell p(x|\theta_\ell)}{p(x|\Theta)} + \alpha_\ell \frac{\partial p(x|\theta_\ell)}{\partial \theta_\ell} - \alpha_\ell \frac{\partial p(x|\theta_\ell)}{\partial \theta_\ell} \right) dx
$$

$$
= -\int \alpha_\ell (t(x) - \phi(\theta_\ell)) p(x|\theta_\ell) \log p(c_\ell|x;\Theta) dx
$$

$$
= -\int (t(x) - \phi(\theta_\ell)) p(c_\ell|x;\Theta) \log p(c_\ell|x;\Theta) p(x|\Theta) dx
$$

Since $\frac{\partial Q(\Theta|\Theta^{(t)})}{\partial \theta_k} = \sum_{i=1}^{n} t(x_i)p(c_\ell|x_i;\Theta^{(t)}) - \phi(\theta_\ell)\sum_{i=1}^{n} p(c_\ell|x_i;\Theta^{(t)})$, we have

$$
\begin{aligned}
\frac{\partial \widetilde{Q}(\Theta|\Theta^{(t)})}{\partial \theta_k} &= \frac{\partial Q(\Theta|\Theta^{(t)})}{\partial \theta_k} - \gamma \frac{\partial H(C|X;\Theta)}{\partial \theta_k} \\
&= \sum_{i=1}^{n} t(x_i)p(c_\ell|x_i;\Theta^{(t)}) + \gamma \int t(x)p(c_\ell|x;\Theta)\log p(c_\ell|x;\Theta)p(x|\Theta)dx \\
&\quad - \phi(\theta_\ell)\left(\sum_{i=1}^{n} p(c_\ell|x_i;\Theta^{(t)}) + \gamma \int p(c_\ell|x;\Theta)\log p(c_\ell|x;\Theta)p(x|\Theta)dx\right)
\end{aligned}
$$

Equating it to zero, we get

$$
\phi(\theta_\ell) = \frac{\sum_{i=1}^{n} t(x_i)p(c_\ell|x_i;\Theta^{(t)}) + \gamma \int t(x)p(c_\ell|x;\Theta)\log p(c_\ell|x;\Theta)p(x|\Theta)dx}{\sum_{i=1}^{n} p(c_\ell|x_i;\Theta^{(t)}) + \gamma \int p(c_\ell|x;\Theta)\log p(c_\ell|x;\Theta)p(x|\Theta)dx}
$$

Bibliography

[Agarwal and Bafna, 1998] Agarwal, P. K. and Bafna, V. (1998). Detecting non-adjoining correlations with signals in DNA. In *Proceedings of the 2nd Annual International Conference on Research in Computational Molecular Biology*, pages 1–7.

[Aggarwal et al., 2006] Aggarwal, A., Guo, D. L., Hoshida, Y., Yuen, S. T., Chu, K.-M., So, S., Boussioutas, A., Chen, X., Bowtell, D., Aburatani, H., Leung, S. Y., and Tan, P. (2006). Topological and functional discovery in a gene coexpression meta-network of gastric cancer. *Cancer Research*, 66(1):232–241.

[Aggarwal and Yu, 2001] Aggarwal, C. C. and Yu, P. S. (2001). Outlier detection for high dimensional data. In *Proceedings of ACM SIGMOD International Conference on Management of Data*, pages 37–46.

[Agrawal et al., 1998] Agrawal, R., Gehrke, J., Gunopolos, D., and Raghavan, P. (1998). Automatic subspace clustering of high dimensional data for data mining applications. In *Proceedings of the ACM SIGMOD International Conference on Management of Data*, pages 94–105.

[Aizerman et al., 1964] Aizerman, M., E.Braverman, and L.Rozonoer (1964). Theoretical foundations of the potential function method in pattern recognition learning. *Automation and Remote Control*, 25:821–837.

[Akaike, 1973] Akaike, H. (1973). Information theory and an extension of the maximum likelihood principle. In Petrov, B. N. and Csaki, F., editors, *Second International Symposium on Information Theory*, pages 267–281.

[Alizadeh et al., 2000] Alizadeh, A. A., Eisen, E. B., Davis, R. E., Ma, C., Lossos, I. S., Rosenwald, A., Boldrick, J. C., Sabet, H., Tran, T., Yu, X., Powell, J. I., Yang, L., Marti, G. E., Moore, T., Hudson Jr., J., Lu, L., Lewis, D. B., Tibshirani, R., Sherlock, G., Chan, W. C., Greiner, T. C., Weisenburger, D. D., Armitage, J. O., Warnke, R., Levy, R., Wilson, W., Grever, M. R., Byrd, J. C., Botstein, D., Brown, P. O., and Staudt, L. M. (2000). Distinct types of diffuse large B-cell lymphoma identified by gene expression profiling. *Nature*, 403(6769):503–511.

[Alon et al., 1999] Alon, U., Barkai, N., Notterman, D. A., Gish, K., Ybarra, S., Mack, D., and Levine, A. J. (1999). Broad patterns of gene expression revealed by clustering analysis of tumor and normal colon tissues probed by oligonucleotide array. *Proceedings of the National Academy of Sciences of USA*, 96(12):6745–6750.

[Altschul, 1991] Altschul, S. F. (1991). Amino acid substitution matrices from an information theoretic perspective. *Journal of Molecular Biology*, 219(3):555–565.

[Ankerst et al., 1999] Ankerst, M., Breunig, M. M., Kriegel, H.-P., and Sander, J. (1999). OPTICS: Ordering points to identify the clustering structure. In *Proceedings of ACM SIGMOD International Conference on Management of Data*, pages 49–60.

[Armbrust et al., 2004] Armbrust, E. V., Berges, J. A., and *et al.* (2004). The genome of the diatom Thalassiosira Pseudonana: Ecology, evolution, and metabolism. *Science*, 306(5693):79–86.

[Arning et al., 1996] Arning, A., Agrawal, R., and Raghavan, P. (1996). A linear method for deviation detection in large databases. In Simoudis, E., Han, J., and Fayyad, U., editors, *Proceedings of the 2nd International Conference on Knowledge Discovery and Data Mining*, pages 164–169.

[Aronszajn, 1950] Aronszajn, N. (1950). Theory of reproducing kernels. *Transactions of the American Mathematical Society*, 68(3):337–404.

[Bader and Hogue, 2003] Bader, G. D. and Hogue, C. W. (2003). An automated method for finding molecular complexes in large protein interaction networks. *BMC Bioinformatics*, 4(2).

192

[BALL and HALL, 1965] BALL, G. H. and HALL, D. J. (1965). ISO-DATA: A novel method of data analysis and classification. Technical report, Stanford University, Stanford, CA.

[Barbara et al., 2002] Barbara, D., Couto, J., and Li, Y. (2002). COOL-CAT: an entropy-based algorithm for categorical clustering. In *Proceedings of the 11th International Conference on Information and Knowledge Management*, pages 582–589.

[Barndorff-Nielsen, 1978] Barndorff-Nielsen, O. E. (1978). *Information and Exponential Families in Statistical Theory*. John Wiley & Sons, New York.

[Barnett and Lewis, 1994] Barnett, V. and Lewis, T. (1994). *Outliers in Statistical Data*. John Wiley & Sons, Chichester, 3rd edition.

[Bay and Schwabacher, 2003] Bay, S. D. and Schwabacher, M. (2003). Mining distance-based outliers in near linear time with randomization and a simple pruning rule. In *Proceedings of the 9th ACM SIGKDD International Conference on Knowledge Discovery and Data Mining*, pages 29–38.

[Beer and Tavazoie, 2004] Beer, M. A. and Tavazoie, S. (2004). Predicting gene expression from sequence. *Cell*, 117:185–198.

[Belhumeur et al., 1997] Belhumeur, P. N., Hespanha, J., and Kriegman, D. J. (1997). Eigenfaces vs. fisherfaces: Recognition using class specific linear projection. *IEEE Transactions on Pattern Analysis and Machine Intelligence*, 19(7):711–720.

[Bellman, 1961] Bellman, R. E. (1961). *Adaptive Control Precesses: A Guided Tour*. Princeton University Press, Princeton, NJ.

[Ben-Dor et al., 2000] Ben-Dor, A., Bruhn, L., Friedman, N., Nachman, I., Schummer, M., and Yakhini, Z. (2000). Tissue classification with gene expression profiles. In *Proceedings of the 4th Annual International Conference on Research in Computational Molecular Biology*, pages 54–64.

[Ben-Dor et al., 2002] Ben-Dor, A., Chor, B., Karp, R., and Yakhini, Z. (2002). Discovering local structure in gene expression data: the order-preserving submatrix problem. In *Proceedings of the 6th Annual*

International Conference on Research in Computational Molecular Biology, pages 49–57.

[Berg et al., 1984] Berg, C., Christensen, J. P. R., and Ressel, P. (1984). *Harmonic Analysis on Semigroups.* Springer-Verlag.

[Berghel and Roach, 1996] Berghel, H. and Roach, D. (1996). An extension of Ukkonen's enhanced dynamic programming ASM algorithm. *ACM Transactions on Information Systems*, 14(1):94–106.

[Bergmann et al., 2004] Bergmann, S., Ihmels, J., and Barkai, N. (2004). Similarities and differences in genome-wide expression data of six organisms. *PLoS Biology*, 2:E9.

[Bertsekas, 1999] Bertsekas, D. P. (1999). *Nonlinear Programming.* Athena Scientific, Belmont, MA, 2nd edition.

[Bharadwaj et al., 2004] Bharadwaj, R., Qi, W., and Yu, H. (2004). Identification of two novel components of the human ndc80 kinetochore complex. *J. Biol. Chem.*, 279(13):13076–13085.

[Bishop, 1995] Bishop, C. M. (1995). *Neural Networks for Pattern Recognition.* Oxford University Press, New York.

[Bø and Jonassen, 2002] Bø, T. H. and Jonassen, I. (2002). New feature subset selection procedures for classification of expression profiles. *Genome Biology*, 3(4):1–11.

[Bolstad et al., 2003] Bolstad, B., Irizarry, R. A., Astrand, M., and Speed, T. P. (2003). A comparison of normalization methods for high density oligonucleotide array data based on bias and variance. *Bioinformatics*, 19:185–193.

[Breiman, 2001] Breiman, L. (2001). Random forests. *Machine Learning*, 45(1):5–32.

[Breunig et al., 2000] Breunig, M. M., Kriegel, H.-P., Ng, R. T., and Sander, J. (2000). LOF: Identifying density-based local outliers. In *Proceedings of ACM SIGMOD International Conference on Management of Data*, pages 93–104.

[Burge and Karlin, 1997] Burge, C. and Karlin, S. (1997). Prediction of complete gene structures in human genomic DNA. *Journal of Molecular Biology*, 268(1):78–94.

[Calinski and Harabasz, 1974] Calinski, R. B. and Harabasz, J. (1974). A dendrite method for cluster analysis. *Communications in Statistics*, 3:1–27.

[Cassimeris and Morabito, 2004] Cassimeris, L. and Morabito, J. (2004). TOGp, the human homolog of XMAP215/Dis1, is required for centrosome integrity, spindle pole organization, and bipolar spindle assembly. *Mol. Biol. Cell*, 15:1580–1590.

[Cheeseman and Stutz, 1995] Cheeseman, P. and Stutz, J. (1995). Bayesian classification (AutoClass): Theory and results. In *Advances in Knowledge Discovery and Data Mining*, pages 153–180, Menlo Park, CA. AAAI Press.

[Chen and Yuan, 2006] Chen, J. and Yuan, B. (2006). Detecting functional modules in the yeast protein-protein interaction network. *Bioinformatics*, 22:2283–2290.

[Chen et al., 2000] Chen, L.-F., Liao, H.-Y. M., Ko, M.-T., Lin, J.-C., and Yu, G.-J. (2000). A new LDA-based face recognition system which can solve the small sample size problem. *Pattern Recognition*, 33(10):1713–1726.

[Cheng and Church, 2000] Cheng, Y. and Church, G. M. (2000). Biclustering of expression data. In *Proceedings of the 8th International Conference on Intelligent Systems for Molecular Biology*, pages 93–103.

[Chiu, 1996] Chiu, S.-T. (1996). A comparative review of bandwidth selection for kernel density estimation. *Statistica Sinica*, 6:129–145.

[Chretien and Hero, 2000] Chretien, S. and Hero, A. O. (2000). Kullback proximal algorithms for maximum likelihood estimation. *IEEE Transactions on Information Theory*, 46(5):1800–1810.

[Chun et al., 2004] Chun, J. H., Kim, H. K., Kim, E., Kim, I.-H., Kim, J. H., Chang, H. J., Choi, I. J., Lim, H.-S., Kim, I.-J., Kang, H. C., Park, J.-H., Bae, J.-M., and Park, J.-G. (2004). Increased expression of metallothionein is associated with irinotecan resistance in gastric cancer. *Cancer Research*, 64:4703–4706.

[Chung et al., 2006] Chung, M. J., Hogstrand, C., and Lee, S. J. (2006). Cytotoxicity of nitric oxide is alleviated by zinc-mediated expression of antioxidant genes. *Exp. Biol. Med.*, 231:1555–1563.

[Churchill et al., 1989] Churchill, G. A., Daniels, D. L., and Waterman, M. S. (1989). The distribution of restriction enzyme sites *escherichia coli*. *Nucleic Acids Research*, 18(3):589–597.

[Cohen, 1991] Cohen, A. C. (1991). *Truncated and censored samples: theory and applications*. CRC Press.

[Collins et al., 2003] Collins, F. S., Green, E. D., Guttmacher, A. E., and Guyer, M. S. (2003). A vision for the future of genomics research. *Nature*, 422:835–847.

[Cortes et al., 2002] Cortes, C., Haffner, P., and Mohri, M. (2002). Rational kernels. In *Advances in Neural Information Processing Systems 15*, pages 41–56.

[Cortes et al., 2003] Cortes, C., Haffner, P., and Mohri, M. (2003). Positive definite rational kernels. In *Proceedings of 16th Annual Conference on Computational Learning Theory*, pages 41–56.

[Cover and Hart, 1967] Cover, T. M. and Hart, P. E. (1967). Nearest neighbor pattern classification. *IEEE Transactions on Information Theory*, 13(1):21–27.

[Cover and Thomas, 1991] Cover, T. M. and Thomas, J. A. (1991). *Elements of Information Theory*. John Wiley & Sons, New York.

[Cristianini et al., 2001] Cristianini, N., Shawe-Taylor, J., and Kandola, J. (2001). Spectral kernel methods for clustering. In *Advances in Neural Information Processing Systems 14*, pages 649–655.

[Crooks et al., 2004] Crooks, G. E., Hon, G., Chandonia, J. M., and Brenner, S. E. (2004). WebLogo: A sequence logo generator. *Genome Research*, 14:1188–1190.

[Date and Marcotte, 2003] Date, S. V. and Marcotte, E. M. (2003). Discovery of uncharacterized cellular systems by genome-wide analysis of functional linkages. *Nature Biotechnology*, 21:1055–1062.

196

[Davuluri et al., 2001] Davuluri, R., Grosse, I., and Zhang, M. Q. (2001). Computational identification of promoters and first exons in the human genome. *Nature Genetics*, 29(4):412–417.

[Dayhoff et al., 1978] Dayhoff, M. O., Schwartz, R. M., and Orcutt, B. C. (1978). A model of evolutionary change in proteins. In Dayhoff, M. O., editor, *Atlas of Protein Sequence and Structure*, volume 5, pages 345–352.

[Dempster et al., 1977] Dempster, A. P., Laird, N. M., and Rubin, D. B. (1977). Maximum likelihood from incomplete data via the EM algorithm. *Journal of the Royal Statistical Society. Series B*, 39(1):1–38.

[Dettling, 2004] Dettling, M. (2004). BagBoosting for tumor classification with gene expression data. *Bioinformatics*, 20(18):3583–3593.

[Dettling and Bühlmann, 2003] Dettling, M. and Bühlmann, P. (2003). Boosting for tumor classification with gene expression data. *Bioinformatics*, 19(9):1061–1069.

[Dimalanta et al., 2004] Dimalanta, E. T., Lim, A., Runnheim, R., Lamers, C., Churas, C., Forrest, D. K., de Pablo, J. J., Graham, M. D., Coppersmith, S. N., Goldstein, S., and Schwartz, D. C. (2004). A microfluidic system for large DNA molecule arrays. *Anal Chem.*, 76(18):5293–5301.

[Drysdale et al., 2005] Drysdale, R. A., Crosby, M. A., and *et al.* (2005). FlyBase: genes and gene models. *Nucleic Acids Research*, 33:D390–395.

[Duda and Hart, 1973] Duda, R. O. and Hart, P. E. (1973). *Pattern Classification and Scene Analysis*. John Wiley & Sons, New York.

[Dudoit and Fridlyand, 2002] Dudoit, S. and Fridlyand, J. (2002). A prediction-based resampling method for estimating the number of clusters in a dataset. *Genome Biology*, 3(7):research0036.1–0036.21.

[Dudoit et al., 2002] Dudoit, S., Fridlyand, J., and Speed, T. P. (2002). Comparison of discrimination methods for the classification of tumors using gene expression data. *Journal of the American Statistical Association*, 97(457):77–87.

197

[Durbin et al., 1998] Durbin, R., Eddy, S. R., Krogh, A., and Mitchison, G. (1998). *Biological Sequence Analysis: Probabilistic Models of Proteins and Nucleic Acids*. Cambridge University Press.

[Edgar et al., 2002] Edgar, R., Domrachev, M., and Lash, A. E. (2002). Gene expression omnibus: NCBI gene expression and hybridization array data repository. *Nucleic Acids Research*, 30:207–210.

[Eisen et al., 1998] Eisen, M. B., Spellman, P. T., Browndagger, P. O., and Botstein, D. (1998). Cluster analysis and display of genome-wide expression patterns. *Proceedings of the National Academy of Sciences of USA*, 95(25):14863–14868.

[Ester et al., 1996] Ester, M., Kriegel, H.-P., Sander, J., and Xu, X. (1996). A density-based algorithm for discovering clusters in large spatial databases with noise. In Simoudis, E., Han, J., and Fayyad, U., editors, *Proceedings of the 2nd International Conference on Knowledge Discovery and Data Mining*, pages 226–231.

[Everitt et al., 2001] Everitt, B. S., Landau, S., and Leese, M. (2001). *Cluster analysis*. Oxford University Press, New York, 4th edition.

[Figueiredo and Jain, 2002] Figueiredo, M. A. T. and Jain, A. K. (2002). Unsupervised learning of finite mixture models. *IEEE Transactions on Pattern Analysis and Machine Intelligence*, 24(3):381–396.

[Fisher, 1936] Fisher, R. A. (1936). The use of multiple measurements in taxonomic problems. *Annual of Eugenics*, 7:179–188.

[Fix and Hodges, 1951] Fix, E. and Hodges, J. (1951). Discriminatory analysis, nonparametric discrimination: consistency properties. Technical report, USAF School of Aviation Medicine, Randolph Field, TX.

[Flannick et al., 2006] Flannick, J., Novak, A., Srinivasan, B. S., McAdams, H. H., and Batzoglou, S. (2006). Graemlin: general and robust alignment of multiple large interaction networks. *Genome Research*, 16:1169–1181.

[Foley, 1972] Foley, D. H. (1972). Considerations of sample and feature size. *IEEE Transactions on Information Theory*, 18(5):618–626.

[Fraley and Raftery, 2002] Fraley, C. and Raftery, A. E. (2002). MCLUST: Software for model-based clustering, discriminant analysis and density estimation. Technical Report 415R, Department of Statistics, University of Washington.

[Freifelder, 1987] Freifelder, D. (1987). *Molecular Biology*. Jones and Bartlett, Boston, MA, 2nd edition.

[Friedman, 1989] Friedman, J. H. (1989). Regularized discriminant analysis. *Journal of the American Statistical Association*, 84:165–175.

[Fukunaga, 1990] Fukunaga, K. (1990). *Introduction to Statistical Pattern Recognition*. Academic Press, New York, 2nd edition.

[Furey et al., 2000] Furey, T. S., Cristianini, N., Duffy, N., Bednarski, D. W., Schummer, M., and Haussler, D. (2000). Support vector machine classification and validation of cancer tissue samples using microarray expression data. *Bioinformatics*, 16(10):906–914.

[Gasch and Eisen, 2002] Gasch, A. P. and Eisen, M. B. (2002). Exploring the conditional coregulation of yeast gene expression through fuzzy k-means clustering. *Genome Biology*, 3(11):1–22.

[Geman and Geman, 1984] Geman, S. and Geman, D. (1984). Stochastic relaxation, Gibbs distribution and the Bayesian restoration of images. *IEEE Transactions on Pattern Analysis and Machine Intelligence*, 6(6):721–741.

[Getz et al., 2000] Getz, G., Levine, E., and Domany, E. (2000). Coupled two-way clustering analysis of gene microarray data. *Proceedings of the National Academy of Sciences of USA*, 97(22):12079–12084.

[Golchin and Paliwal, 1997] Golchin, F. and Paliwal, K. K. (1997). Minimum-entropy clustering and its application to lossless image coding. In *Proceedings of IEEE International Conference on Image Processing*, volume II, pages 262–265.

[Gollub et al., 2003] Gollub, J., Ball, C. A., Binkley, G., Demeter, J., Finkelstein, D. B., Hebert, J. M., Hernandez-Boussard, T., Jin, H., Kaloper, M., Matese, J. C., Schroeder, M., Brown, P. O., Botstein, D., and Sherlock, G. (2003). The Stanford Microarray Database: data access and quality assessment tools. *Nucleic Acids Research*, 31:94–96.

[Golub et al., 1999] Golub, T. R., Slonim, D. K., Tamayo, P., Huard, C., Gaasenbeek, M., Mesirov, J. P., Coller, H., Loh, M. L., Downing, J. R., Caligiuri, M. A., Bloomfield, C. D., and Lander, E. S. (1999). Molecular classification of cancer: Class discovery and class prediction by gene expression monitoring. *Science*, 536:531–537.

[Gordon, 1999] Gordon, A. (1999). *Classification*. Chapman & Hall, London, 2nd edition.

[Grahne and Zhu, 2003] Grahne, G. and Zhu, J. (2003). Efficiently using prefix-trees in mining frequent itemsets. *Proceedings of the ICDM Workshop on Frequent Itemset Mining Implementations*.

[Grimmett and Stirzaker, 1982] Grimmett, G. and Stirzaker, D. (1982). *Probability and Random Processes*. Oxford University Press.

[Guha et al., 1998] Guha, S., Rastogi, R., and Shim, K. (1998). CURE: An efficient clustering algorithm for large databases. In Haas, L. M. and Tiwary, A., editors, *Proceedings of ACM SIGMOD International Conference on Management of Data*, pages 73–84.

[Hampel et al., 1986] Hampel, F. R., Ronchetti, E. M., Rousseeuw, P. J., and Stahel, W. A. (1986). *Robust Statistics: The Approach Based on Influence Functions*. Wiley, New York.

[Han and Kamber, 2000] Han, J. and Kamber, M. (2000). *Data Mining: Concepts and Techniques*. Morgan Kaufmann Publishers, San Francisco.

[Hansen and Yu, 2001] Hansen, M. H. and Yu, B. (2001). Model selection and the principle of minimum description length. *Journal of the American Statistical Association*, 96(454):746–774.

[Hartigan, 1975] Hartigan, J. A. (1975). *Clustering Algorithms*. John Wiley & Sons, New York.

[Hastie and Tibshirani, 1995] Hastie, T. and Tibshirani, R. (1995). Penalized discriminant analysis. *The Annals of Statistics*, 23:73–102.

[Hastie et al., 1994] Hastie, T., Tibshirani, R., and Buja, A. (1994). Flexible discriminant analysis by optimal scoring. *Journal of the American Statistical Association*, 89:1255–1270.

[Hatzigeorgiou, 2002] Hatzigeorgiou, A. G. (2002). Translation initiation start prediction in human cDNAs with high accuracy. *Bioinformatics*, 18(2):343–350.

[Havrda and Charvat, 1967] Havrda, J. and Charvat, F. (1967). Quantification method of classification processes: Concept of structural α-entropy. *Kybernetika*, 3:30–35.

[Hawkins, 1980] Hawkins, D. (1980). *Identification of Outliers*. Chapman & Hall, London.

[Hawkins et al., 2002] Hawkins, S., He, H., Williams, G., and Baxter, R. (2002). Outlier detection using replicator neural networks. In *Proceedings of the 4th International Conference on Data Warehousing and Knowledge Discovery*, pages 170–180.

[Henikoff and Henikoff, 1992] Henikoff, S. and Henikoff, J. G. (1992). Amino acid substitution matrices from protein blocks. *Proceedings of the National Academy of Sciences of USA*, 89(22):10915–10919.

[Hermjakob et al., 2004] Hermjakob, H., Montecchi-Palazzi, L., Lewington, C., Mudali, S., Kerrien, S., Orchard, S., Vingron, M., Roechert, B., Roepstorff, P., Valencia, A., Margalit, H., Armstrong, J., Bairoch, A., Cesareni, G., Sherman, D., and Apweiler, R. (2004). Intact: an open source molecular interaction database. *Nucleic Acids Research*, 32:D452–455.

[Hinneburg and Keim, 1998] Hinneburg, A. and Keim, D. A. (1998). An efficient approach to clustering in large multimedia databases with noise. In *Proceedings of the 4th International Conference on Knowledge Discovery and Data Mining*, pages 58–64.

[Hinnebusch, 1997] Hinnebusch, A. G. (1997). Translational regulation of yeast GCN4. *Journal of Biological Chemistry*, 272(35):21661–21664.

[Hong and Yang, 1991] Hong, Z.-Q. and Yang, J.-Y. (1991). Optimal discriminant plane for a small number of samples and design method of classifier on the plane. *Pattern Recognition*, 24(4):317–324.

[Huang et al., 2002] Huang, R., Liu, Q., Lu, H., and Ma, S. (2002). Solving the small sample size problem of LDA. In *Proceedings of 16th International Conference on Pattern Recognition*, volume 3, pages 29–32.

[Huber, 1981] Huber, P. J. (1981). *Robust Statistics*. Wiley, New York.

[Hubert and Arabie, 1985] Hubert, L. and Arabie, P. (1985). Comparing partitions. *Journal of Classification*, 2:193–218.

[Hughes et al., 2000] Hughes, T. R., Marton, M. J., Jones, A. R., Roberts, C. J., Stoughton, R., Armour, C. D., Bennett, H. A., Coffey, E., Dai, H., He, Y. D., Kidd, M. J., King, A. M., Meyer, M. R., Slade, D., Lum, P. Y., Stepaniants, S. B., Shoemaker, D. D., Gachotte, D., Chakraburtty, K., Simon, J., Bard, M., and Friend, S. H. (2000). Functional discovery via a compendium of expression profile. *Cell*, 102(1):109–126.

[Hwang et al., 2006] Hwang, W., Cho, Y. R., Zhang, A., and Ramanathan, M. (2006). A novel functional module detection algorithm for protein-protein interaction networks. *Algorithms for Molecular Biology*, 1(24).

[Ideker et al., 2001] Ideker, T., Thorsson, V., Ranish, J. A., Christmas, R., Buhler, J., Eng, J. K., Bumgarner, R., Goodlett, D. R., Aebersold, R., and Hood, L. (2001). Integrated genomic and proteomic analyses of a systematically perturbed metabolic network. *Science*, 292:929–934.

[International Human Genome Sequencing Consortium, 2001] International Human Genome Sequencing Consortium (2001). Initial sequencing and analysis of the human genome. *Nature*, 409:860–921.

[Izmailova et al., 2003] Izmailova, E., Bertley, F. M., Huang, Q., Makori, N., Miller, C. J., Young, R. A., and Aldovini, A. (2003). HIV-1 Tat reprograms immature dendritic cells to express chemoattractants for activated T cells and macrophages. *Nature Medicine*, 9:191–197.

[Jain and Chandrasekaran, 1982] Jain, A. K. and Chandrasekaran, B. (1982). Dimensionality and sample size considerations in pattern recognition practice. In Krishnaiah, P. and Kanal, L., editors, *Handbook of Statistics*, volume 2, pages 835–855. Amsterdam, North Holland.

[Jain and Dubes, 1988] Jain, A. K. and Dubes, R. C. (1988). *Algorithms for Clustering Data*. Prentice Hall, Englewood Cliffs, NJ.

202

[Jain et al., 1999] Jain, A. K., Murty, M. N., and Flyn, P. J. (1999). Data clustering: A review. *ACM Computing Surveys*, 31(3):264–323.

[Jeffery, 2003a] Jeffery, C. J. (2003a). Moonlighting proteins: old proteins learning new tricks. *Trends in Genetics*, 19:415–417.

[Jeffery, 2003b] Jeffery, C. J. (2003b). Multifunctional proteins: examples of gene sharing. *Annals of Medicine*, 35:28–35.

[Jolliffe, 1986] Jolliffe, I. T. (1986). *Principal Component Analysis*. Springer-Verlag, New York.

[Kannan et al., 2000] Kannan, R., Vempala, S., and Vetta, A. (2000). On clusterings: good, bad and spectral. In *Proceedings of the 41th Annual Symposium on Foundations of Computer Science*, pages 367–377.

[Kapur, 1967] Kapur, J. N. (1967). Generalised entropy of order α and type β. *The Mathematics Seminar*, 4:78–94.

[Kapur, 1994] Kapur, J. N. (1994). *Measures of information and their applications*. John Wiley & Sons, New York.

[Kelley et al., 2003] Kelley, B. P., Sharan, R., Karp, R. M., Sittler, T., Root, D. E., Stockwell, B. R., and Ideker, T. (2003). Conserved pathways within bacteria and yeast as revealed by global protein network alignment. *Proc. Natl. Acad. Sci. USA*, 100:11394–11399.

[Khan et al., 2001] Khan, J., Wei, J. S., Ringnér, M., Saal, L. H., Ladanyi, M., Westermann, F., Berthold, F., Schwab, M., Antonescu, C. R., Peterson, C., and Meltzer, P. S. (2001). Classification and diagnostic prediction of cancers using gene expression profiling and artificial neural networks. *Nature Medicine*, 7(6):673–679.

[Kimura, 1980] Kimura, M. (1980). A simple method for estimating evolutionary rates of base substitutions through comparative studies of nucleotide sequences. *Journal of Molecular Evolution*, 16(2):111–120.

[Kirkpatrick et al., 1983] Kirkpatrick, S., Gelatt, C. D., and Vecchi, M. P. (1983). Optimization by simulated annealing. *Science*, 220:671–680.

[Kluger et al., 2003] Kluger, Y., Basri, R., Chang, J. T., and Gerstein, M. (2003). Spectral biclustering of microarray data: coclustering genes and conditions. *Genome Research*, 13(4):703–716.

[Knorr et al., 2000] Knorr, E. M., Ng, R. T., and Tucakov, V. (2000). Distance-based outliers: Algorithms and applications. *The VLDB Journal*, 8:237–253.

[Kohonen, 2001] Kohonen, T. (2001). *Self-Organizing Maps*. Springer-Verlag, New York, 3rd edition.

[Kolmogorov, 1965] Kolmogorov, A. N. (1965). Three approaches for defining the concept of information quantity. *Information Transmission*, 1:3–11.

[Kolmogorov, 1968] Kolmogorov, A. N. (1968). Logical basis for information theory and probability theory. *IEEE Transactions on Information Theory*, 14(5):662–664.

[Kolmogorov, 1983a] Kolmogorov, A. N. (1983a). Combinatorial foundations of information theory and the calculus of probabilities. *Russian Mathematical Surveys*, 38:29–40.

[Kolmogorov, 1983b] Kolmogorov, A. N. (1983b). *On logical foundations of probability theory*, volume 1021 of *Lecture Notes in Mathematics*, pages 1–5. Springer, New York.

[Koyuturk et al., 2006] Koyuturk, M., Kim, Y., Subramaniam, S., Szpankowski, W., and Grama, A. (2006). Detecting conserved interaction patterns in biological networks. *J Comput Biol*, 13:1299–1322.

[Kozak, 1987a] Kozak, M. (1987a). An analysis of 5′-noncoding sequences from 699 vertebrate messenger RNAs. *Nucleic Acids Research*, 15(20):8125–8148.

[Kozak, 1987b] Kozak, M. (1987b). At least six nucleotides preceding the AUG initiator codon enhance translation in mammalian cells. *Journal of Molecular Biology*, 196(4):947–950.

[Kozak, 1987c] Kozak, M. (1987c). Effects of intercistronic length on the efficiency of reinitiation by eucaryotic ribosomes. *Molecular and Cellular Biology*, 7(10):3438–3445.

[Kozak, 1989] Kozak, M. (1989). The scanning model for translation: An update. *Journal of Cell Biology*, 108:229–241.

[Kozak, 1996] Kozak, M. (1996). Interpreting cDNA sequences: Some insights from studies on translation. *Mamalian Genome*, 7(8):563–574.

[Kozak, 1999] Kozak, M. (1999). Initiation of translation in prokaryotes and eukaryotes. *Gene*, 234(2):187–208.

[Lazzeroni and Owen, 2002] Lazzeroni, L. and Owen, A. (2002). Plaid models for gene expression data. *Statistica Sinica*, 12(1):61–85.

[Lee and Choi, 2005] Lee, Y. and Choi, S. (2005). Maximum within-cluster association. *Pattern Recognition Letters*, 26(10):1412–1422.

[Levenshtein, 1966] Levenshtein, V. (1966). Binary codes capable of correcting deletions, insertions and reversals. *Soviet Physics Daklady*, 10:707–710.

[Li et al., 2005] Li, H., Zhang, K., and Jiang, T. (2005). The regularized EM algorithm. In *Proceedings of the 20th National Conference on Artifical Intelligence*, pages 807–812.

[Li et al., 2001a] Li, L., Darden, T. A., Weinberg, C. R., Levine, A. J., and Pedersen, L. G. (2001a). Gene assessment and sample classification for gene expression data using a genetic algorithm/k-nearest neighbor method. *Combinatorial Chemistry & High Throughput Screening*, 4(8):727–739.

[Li et al., 2001b] Li, L., Weinberg, C. R., Darden, T. A., and Pedersen, L. G. (2001b). Gene selection for sample classification based on gene expression data: study of sensitivity to choice of parameters of the GA/KNN method. *Bioinformatics*, 17(12):1131–1142.

[Li, 2003] Li, L. M. (2003). Blind inversion needs distribution (BIND): The general notion and case studies. In Goldstein, D., editor, *Science and Statistics: A Festschrift for Terry Speed*, volume 40 of *IMS Lecture Note Series*, pages 273–293.

[Li and Vitányi, 1997] Li, M. and Vitányi, P. (1997). *An Introduction to Kolmogorov Complexity and its Applications*. Springer-Verlag, New York, 2nd edition.

[Li et al., 2004] Li, T., Zhang, C., and Ogihara, M. (2004). A comparative study of feature selection and multiclass classification methods for tissue classification based on gene expression. *Bioinformatics*, 20(15):2429–2437.

[Li and Graur, 1991] Li, W.-H. and Graur, D. (1991). *Fundamentals of Molecular Evolution*. Sinauer Associates, Sunderland, MA.

[Lin et al., 1999] Lin, J., Qi, R., Aston, C., Jing, J., Anantharaman, T. S., Mishra, B., White, O., Daly, M. J., Minton, K. W., Venter, J. C., , and Schwartz, D. C. (1999). Whole-genome shotgun optical mapping of *deinococcus radiodurans*. *Science*, 285(5433):1558–1562.

[Liu et al., 2004] Liu, J., Yang, J., and Wang, W. (2004). Biclustering of gene expression data by tendency. In *Proceedings of the 3rd IEEE Computational Systems Bioinformatics Conference*, pages 182–193.

[Liu, 1996] Liu, J. S. (1996). Peskun's theorem and a modified discrete-state gibbs sampler. *Biometrika*, 83(3):681–682.

[Lockhart et al., 1996] Lockhart, D. J., Dong, H., Byrne, M. C., Follettie, M. T., Gallo, M. V., Chee, M. S., Mittmann, M., Wang, C., Kobayashi, M., Horton, H., and Brown, E. L. (1996). Expression monitoring by hybridization to high-density oligonucleotide arrays. *Nature Biotechnology*, 14(13):1675–1680.

[Loftsgaarden and Quesenberry, 1965] Loftsgaarden, D. O. and Quesenberry, C. P. (1965). A nonparametric estimate of a multivariate density function. *Annals of Mathematical Statistics*, 36(3):1049–1051.

[Luo et al., 2007] Luo, F., Yang, Y., Chen, C. F., Chang, R., Zhou, J., and Scheuermann, R. H. (2007). Modular organization of protein interaction networks. *Bioinformatics*, 23:207–214.

[Lussier et al., 2006] Lussier, Y., Borlawsky, T., Rappaport, D., Liu, Y., and Friedman, C. (2006). PhenoGO: assigning phenotypic context to gene ontology annotations with natural language processing. In *Proceedings of Pacific Symposium on Biocomputing*, pages 64–75.

[Luukkonen et al., 1995] Luukkonen, B. G. M., Tan, W., and Schwartz, S. (1995). Efficiency of reinitiation of translation on human immunodeficiency virus type 1 mRNAs is determined by the length of the upstream open reading frame and by intercistronic distance. *Journal of Virology*, 69(7):4086–4094.

[Madeira and Oliveira, 2004] Madeira, S. C. and Oliveira, A. L. (2004). Biclustering algorithms for biological data analysis: A survey. *IEEE/ACM Transactions on Computational Biology and Bioinformatics*, 1(1):24–45.

[Mewes et al., 1997] Mewes, H. W., Albermann, K., Bahr, M., Frishman, D., Gleissner, A., Hani, J., Heumann, K., Kleine, K., Maierl, A., Oliver, S. G., Pfeiffer, F., and Zollner, A. (1997). Overview of the yeast genome. *Nature*, 387:7–65.

[Milligan and Cooper, 1985] Milligan, G. W. and Cooper, M. C. (1985). An examination of procedures for determining the number of clusters in a data set. *Psychometrika*, 50:159–179.

[Milligan and Cooper, 1986] Milligan, G. W. and Cooper, M. C. (1986). A study of the comparability of external criteria for hierarchical cluster analysis. *Multivariate Behavioral Research*, 21:441–458.

[Mount, 2001] Mount, D. W. (2001). *Bioinformatics: Sequence and Genome Analysis*. Cold Spring Harbor Laboratory Press, Cold Spring Harbor, N.Y.

[Mukherjee et al., 1998] Mukherjee, S., Tamayo, P., Slonim, D., Verri, A., Golub, T., Mesirov, J. P., and Poggio, T. (1998). Support vector machine classification of microarray data. AI Memo 1677, MIT CBCL, Cambridge, MA.

[Murali and Kasif, 2003] Murali, T. M. and Kasif, S. (2003). Extracting conserved gene expression motifs from gene expression data. In *Proceedings of the Pacific Symposium on Biocomputing*, pages 77–88.

[Ng et al., 2001] Ng, A. Y., Jordan, M. I., and Weiss, Y. (2001). On spectral clustering: Analysis and an algorithm. In *Advances in Neural Information Processing Systems 14*, pages 849–856.

[Ng and Han, 1994] Ng, R. T. and Han, J. (1994). Efficient and effective clustering methods for spatial data mining. In Bocca, J., Jarke, M., and Zaniolo, C., editors, *Proceedings of the 20th International Conference on Very Large Data Bases*, pages 144–155.

[Niehrs and Pollet, 1999] Niehrs, C. and Pollet, N. (1999). Synexpression groups in eukaryotes. *Nature*, 402:483–487.

[Okudaira et al., 2005] Okudaira, K., Ohno, K., Yoshida, H., Asano, M., Hirose, F., and Yamaguchi, M. (2005). Transcriptional regulation of the *drosophila* orc2 gene by the DREF pathway. *Biochim. Biophys. Acta.*, 1732:23–30.

[Oldham et al., 2006] Oldham, M. C., Horvath, S., and Geschwind, D. H. (2006). Conservation and evolution of gene coexpression networks in human and chimpanzee brains. *Proc. Natl. Acad. Sci. USA*, 103:17973–17978.

[Parzen, 1962] Parzen, E. (1962). On estimation of a probability density function and mode. *Annals of Mathematical Statistics*, 33(3):1065–1076.

[Pedersen and Nielsen, 1997] Pedersen, A. G. and Nielsen, H. (1997). Neural network prediction of translation initiation sites in eukaryotes: Perspectives for EST and genome analysis. In *Proceedings of the 5th International Conference on Intelligent Systems for Molecular Biology*, pages 226–233.

[Petrovskiy, 2003] Petrovskiy, M. I. (2003). Outlier detection algorithms in data mining systems. *Programming and Computer Software*, 29(4):228–237.

[Platt, 1999] Platt, J. (1999). Fast training of support vector machines using sequential minimal optimization. In Schölkopf, B., Burges, C. J. C., and Smola, A. J., editors, *Advances in Kernel Methods — Support Vector Learning*, pages 185–208, Cambridge, MA. MIT Press.

[Poggio et al., 1985] Poggio, T., Torre, V., and Koch, C. (1985). Computational vision and regularization theory. *Nature*, 317:314–319.

[Pomeroy et al., 2002] Pomeroy, S. L., Tamayo, P., Gaasenbeek, M., Sturla, L. M., Angelo, M., McLaughlin, M. E., Kim, J. Y., Goumnerova, L. C., Black, P. M., Lau, C., Allen, J. C., Zagzag, D., Olson, J. M., Curran, T., Wetmore, C., Biegel, J. A., Poggio, T., Mukherjee, S., Rifkin, R., Califano, A., G, G. S., Louis, D. N., Mesirov, J. P., Lander, E. S., and Golub, T. R. (2002). Prediction of central nervous system embryonal tumour outcome based on gene expression. *Nature*, 415(6870):436–442.

[Pruitt and Maglott, 2001] Pruitt, K. D. and Maglott, D. R. (2001). Refseq and locuslink: Ncbi gene-centered resources. *Nucleic Acids Research*, 29(1):137–140.

[Qian et al., 2001] Qian, J., Dolled-Filhart, M., Lin, J., Yu, H., and Gerstein, M. (2001). Beyond synexpression relationships: local clustering of time-shifted and inverted gene expression profiles identifies new, biologically relevant interactions. *J. Mol. Biol,* 314:1053–1066.

[Ramaswamy et al., 2000] Ramaswamy, S., Rastogi, R., and Shim, K. (2000). Efficient algorithms for mining outliers from large data sets. In *Proceedings of ACM SIGMOD International Conference on Management of Data*, pages 427–438.

[Ramaswamy et al., 2001] Ramaswamy, S., Tamayo, P., Rifkin, R., Mukherjee, S., Yeang, C. H., Angelo, M., Ladd, C., Reich, M., Latulippe, E., Mesirov, J. P., Poggio, T., Gerald, W., Loda, M., Lander, E. S., and Golub, T. R. (2001). Multiclass cancer diagnosis using tumor gene expression signatures. *Proceedings of the National Academy of Science*, 98(26):15149–15154.

[Rand, 1971] Rand, W. M. (1971). Objective criteria for the evaluation of clustering methods. *Journal of American Statistical Association*, 66:846–850.

[Rao, 1948] Rao, C. R. (1948). The utilization of multiple measurements in problems of biological classification. *Journal of the Royal Statistical Society. Series B (Methodological)*, 10:159–203.

[Raudys and Jain, 1991] Raudys, S. J. and Jain, A. K. (1991). Small sample size effects in statistical pattern recognition: Recommendations for practitioners. *IEEE Transactions on Pattern Analysis and Machine Intelligence*, 13(3):252–264.

[Raudys and Pikelis, 1980] Raudys, S. J. and Pikelis, V. (1980). On dimensionality, sample size, classification error, and complexity of classification algorithms in pattern recognition. *IEEE Transactions on Pattern Analysis and Machine Intelligence*, 2:243–251.

[Redner and Walker, 1984] Redner, R. A. and Walker, H. F. (1984). Mixture densities, maximum likelihood and the EM algorithm. *SIAM Review*, 26(2):195–239.

[Renyi, 1961] Renyi, A. (1961). On measures of entropy and information. In *Proceedings of the 4th Berkeley Symposium on Mathematics, Statistics and Probability*, volume 1, pages 547–561. University of California Press.

[Reslewic et al., 2005] Reslewic, S., Zhou, S., Place, M., Zhang, Y., Briska, A., Goldstein, S., Churas, C., Runnheim, R., Forrest, D., Lim, A., Lapidus, A., Han, C. S., Roberts, G. P., Schwartz, D. C., and Walker, H. F. (2005). Whole-genome shotgun optical mapping of *rhodospirillum rubrum*. *Appl. Environ. Microbiol.*, 71(9):5511–5522.

[Rissanen, 1985] Rissanen, J. (1985). Minimum description length principle. In Kotz, S. and Johnson, N. L., editors, *Encyclopedia of Statistical Sciences*, volume 5, pages 523–527. Wiley, New York.

[Roberts et al., 2000] Roberts, S. J., Everson, R. M., and Rezek, I. (2000). Maximum certainty data partitioning. *Pattern Recognition*, 33(5):833–839.

[Roberts et al., 2001] Roberts, S. J., Holmes, C., and Denison, D. (2001). Minimum-entropy data partitioning using reversible jump markov chain monte carlo. *IEEE Transactions on Pattern Analysis and Machine Intelligence*, 23(8):909–914.

[Rosenblatt, 1956] Rosenblatt, M. (1956). Remarks on some nonparametric estimates of a density function. *Annals of Mathematical Statistics*, 27(3):832–837.

[Roth et al., 1998] Roth, F. P., Hughes, J. D., Estep, P. W., and Church, G. M. (1998). Finding dna regulatory motifs within unaligned noncoding sequences clustered by whole-genome mrna quantitation. *Nature Biotechnoly*, 16(10):939–945.

[Ruepp et al., 2004] Ruepp, A., Zollner, A., Maier, D., Albermann, K., Hani, J., Mokrejs, M., Tetko, I., Güldener, U., Mannhaupt, G., Münsterkötter, M., , and Mewes, H. W. (2004). The funcat, a functional annotation scheme for systematic classification of proteins from whole genomes. *Nucleic Acids Research*, 32(18):5539–5545.

[Ruts and Rousseeuw, 1996] Ruts, I. and Rousseeuw, P. (1996). Computing depth contours of bivariate point clouds. *Computer Journal of Computational Statistics and Data Analysis*, 23:153–168.

[Salamov et al., 1998] Salamov, A. A., Nishikawa, T., and Swindells, M. B. (1998). Assessing protein coding region integrity in cDNA sequencing projects. *Bioinformatics*, 14(5):384–390.

[Salzberg, 1997] Salzberg, S. L. (1997). A method for identifying splice sites and translational start sites in eukaryotic mRNA. *Computer Applications in the Biosciences*, 13(4):365–376.

[Schena et al., 1995] Schena, M., Shalon, D., Davis, R. W., and Brown, P. O. (1995). Quantitative monitoring of gene expression patterns with a complementary DNA microarray. *Science*, 270:467–470.

[Schölkopf, 1997] Schölkopf, B. (1997). *Support Vector Learning*. R. Oldenbourg Verlag, Munich.

[Schölkopf et al., 1998] Schölkopf, B., Smola, A., and Müller, K.-R. (1998). Nonlinear component analysis as a kernel eigenvalue problem. *Neural Compution*, 10:1299–1319.

[Schwarz, 1978] Schwarz, G. (1978). Estimating the dimension of a model. *The Annals of Statistics*, 6(2):461–464.

[Scott, 1992] Scott, D. (1992). *Multivariate Density Estimation: Theory, Practice and Visualization*. Wiley, New York.

[Sellers, 1974] Sellers, P. H. (1974). On the theory and computation of evolutionary distances. *SIAM Journal on Applied Mathematics*, 26(4):787–793.

[Shannon, 1948] Shannon, C. E. (1948). A mathematical theory of communication. *Bell System Techical Journal*, 27:379–423 and 623–656.

[Shannon et al., 2003] Shannon, P., Markiel, A., Ozier, O., Baliga, N. S., Wang, J. T., Ramage, D., Amin, N., Schwikowski, B., and Ideker, T. (2003). Cytoscape: a software environment for integrated models of biomolecular interaction networks. *Genome Research*, 13:2498–2504.

[Sharan and Ideker, 2006] Sharan, R. and Ideker, T. (2006). Modeling cellular machinery through biological network comparison. *Nature Biotechnology*, 24:427–433.

[Sharan et al., 2005] Sharan, R., Suthram, S., Kelley, R. M., Kuhn, T., McCuine, S., Uetz, P., Sittler, T., Karp, R. M., and Ideker, T. (2005). Conserved patterns of protein interaction in multiple species. *Proc. Natl. Acad. Sci. USA*, 102:1974–1979.

[Sheikholeslami et al., 1998] Sheikholeslami, G., Chatterjee, S., and Zhang, A. (1998). WaveCluster: A multi-resolution clustering approach for very large spatial databases. In *Proceedings of the 24th International Conference on Very Large Data Bases*, pages 428–439.

[Sheng et al., 2003] Sheng, Q., Moreau, Y., and Moor, B. D. (2003). Biclustering microarray data by Gibbs sampling. *Bioinformatics*, 19(Supplement 2):196–205.

[Singh et al., 2002] Singh, D., Febbo, P. G., Ross, K., Jackson, D. G., Manola, J., Ladd, C., Tamayo, P., Renshaw, A. A., D'Amico, A. V., Richie, J. P., Lander, E. S., Loda, M., Kantoff, P. W., Golub, T. R., and Sellers, W. R. (2002). Gene expression correlates of clinical prostate cancer behavior. *Cancer Cell*, 1(2):203–209.

[Slonim et al., 2000] Slonim, D. K., Tamayo, P., Mesirov, J. P., Golub, T. R., and Lander, E. S. (2000). Class prediction and discovery using gene expression data. In *Proceedings of the 4th Annual International Conference on Research in Computational Molecular Biology*, pages 263–272.

[Solomonoff, 1964] Solomonoff, R. J. (1964). A formal theory of inductive inference. *Information and Control*, 7:1–22 and 224–254.

[Somorjai et al., 2003] Somorjai, R. L., Dolenko, B., and Baumgartner, R. (2003). Class prediction and discovery using gene microarray and proteomics mass spectroscopy data: curses, caveats, cautions. *Bioinformatics*, 19(12):1484–1491.

[Spellman et al., 1998] Spellman, P. T., Sherlock, G., Zhang, M. Q., Iyer, V. R., Anders, K., Eisen, M. B., Brown, P. O., Botstein, D., and

FutcherDagger, B. (1998). Comprehensive identification of cell cycle-regulated genes of the yeast saccharomyces cerevisiae by microarray hybridization. *Molecular Biology of the Cell*, 9(12):3273–3297.

[Spirin and Mirny, 2003] Spirin, V. and Mirny, L. A. (2003). Protein complexes and functional modules in molecular networks. *Proc. Natl. Acad. Sci. USA*, 100:12123–12128.

[Stuart et al., 2003] Stuart, J. M., Segal, E., Koller, D., and Kim, S. K. (2003). A gene-coexpression network for global discovery of conserved genetic modules. *Science*, 302:249–255.

[Su et al., 2003] Su, Y., Murali, T. M., Pavlovic, V., Schaffer, M., and Kasif, S. (2003). RankGene: identification of diagnostic genes based on expression data. *Bioinformatics*, 19(12):1578–1579.

[Tamayo et al., 1999] Tamayo, P., Slonim, D., Mesirov, J., Zhu, Q., Kitareewan, S., Dmitrovsky, E., Lander, E. S., and Golub, T. R. (1999). Interpreting patterns of gene expression with self-organizing maps: methods and application to hematopoietic differentiation. *Proc. Natl. Acad. Sci. USA*, 96:2907–2912.

[Tanay et al., 2002] Tanay, A., Sharan, R., and Shamir, R. (2002). Discovering statistically significant biclusters in gene expression data. *Bioinformatics*, 18(Supplement 1):136–144.

[Tang et al., 2002] Tang, J., Chen, Z., Fu, A. W., and Cheung, D. (2002). Enhancing effectiveness of outlier detections for low density patterns. In *Proceedings of the 6th Pacific-Asia Conference on Knowledge Discovery and Data Mining*, pages 535–548.

[Tavazoie et al., 1999] Tavazoie, S., Hughes, J. D., Campbell, M. J., Cho, R. J., and Church, G. M. (1999). Systematic determination of genetic network architecture. *Nature Genetics*, 22:281–285.

[Tian et al., 1986] Tian, Q., Barbero, M., Gu, Z.-H., and Lee, S. H. (1986). Image classification by the foley-sammon transform. *Optical Engineering*, 25(7):834–840.

[Tibshirani et al., 2001] Tibshirani, R., Walther, G., and Hastie, T. (2001). Estimating the number of clusters in a dataset via the Gap statistic. *Journal of the Royal Statistical Society. Series B (Methodological)*, 63:411–423.

[Tikhonov, 1963] Tikhonov, A. N. (1963). Solution of incorrectly formulated problems and the regularization method. *Soviet Mathematics Doklady*, 4:1035–1038.

[Tornow and Mewes, 2003] Tornow, S. and Mewes, H. (2003). Functional modules by relating protein interaction networks and gene expression. *Nucleic Acids Research*, 31:6283–6289.

[Turk and Pentland, 1991] Turk, M. A. and Pentland, A. P. (1991). Eigenfaces for recognition. *Journal of Cognitive Neuroscience*, 3(1):71–86.

[Turlach, 1993] Turlach, B. A. (1993). Bandwidth selection in kernel density estimation: A review.

[Ukkonen, 1985] Ukkonen, E. (1985). Algorithms for approximate string matching. *Information and Control*, 64:100–118.

[Valouev et al., 2006] Valouev, A., Li, L. M., Liu, Y.-C., Schwartz, D. C., Yang, Y., Zhang, Y., and Waterman, M. S. (2006). Alignment of optical maps. *Journal of Computational Biology*, 13(2):442–462.

[Vapnik, 1995] Vapnik, V. N. (1995). *The Nature of Statistical Learning Thoery*. Springer-Verlag, New York.

[Vapnik, 1998] Vapnik, V. N. (1998). *Statistical Learning Theory*. John Wiley & Sons, New York.

[Vapnik and Chervonenkis, 1974] Vapnik, V. N. and Chervonenkis, A. (1974). *Theory of Pattern Recognition*. Nauka, Moscow.

[Velculescu et al., 1995] Velculescu, V. E., Zhang, L., Vogelstein, B., and Kinzler, K. W. (1995). Serial analysis of gene expression. *Science*, 270:484–487.

[Venter et al., 2001] Venter, J. C., Adams, M. D., and *et al.* (2001). The sequence of the human genome. *Science*, 291:1304–1351.

[Vogt, 2005] Vogt, P. H. (2005). Azoospermia factor (AZF) in Yq11: towards a molecular understanding of its function for human male fertility and spermatogenesis. *Reprod Biomed Online*, 10:81–93.

[Wallace and Boulton, 1968] Wallace, C. S. and Boulton, D. M. (1968). An information measure for classification. *Computer Journal*, 11:185–194.

[Wang et al., 2005] Wang, W., Cherry, J. M., Nochomovitz, Y., Jolly, E., Botstein, D., and Li, H. (2005). Inference of combinatorial regulation in yeast transcriptional networks: a case study of sporulation. *Proc. Natl. Acad. Sci. USA*, 102:1998–2003.

[Wang et al., 1997] Wang, W., Yang, J., and Muntz, R. R. (1997). STING: A statistical information grid approach to spatial data mining. In *Proceedings of the 23th International Conference on Very Large Data Bases*, pages 186–195.

[Waterman, 1995] Waterman, M. S. (1995). *Introduction to Computational Biology*. Chapman and Hall/CRC.

[Waterman et al., 1984] Waterman, M. S., Smith, T. F., and Katcher, H. (1984). Algorithms for restriction map comparisons. *Nucleic Acids Research*, 12:237–242.

[Watson and Crick, 1953] Watson, J. D. and Crick, F. H. C. (1953). Molecular structure of nucleic acids: A structure for deoxyribose nucleic acid. *Nature*, 171:737–738.

[Weiss, 1999] Weiss, Y. (1999). Segmentation using eigenvectors: A unifying view. In *Proceedings of IEEE International Conference on Computer Vision*, pages 975–982.

[Weston et al., 2000] Weston, J., Mukherjee, S., Chapelle, O., Pontil, M., Poggio, T., and Vapnik, V. (2000). Feature selection for SVMs. In Leen, T. K., Dietterich, T., and Tresp, V., editors, *Advances in Neural Information Processing Systems 13*, pages 668–674. MIT Press.

[Wientjes et al., 1993] Wientjes, F. B., Hsuan, J. J., Totty, N. F., and Segal, A. W. (1993). p40phox, a third cytosolic component of the activation complex of the NADPH oxidase to contain src homology 3 domains. *J. Biochem*, 296:557–561.

[Wu et al., 2002] Wu, L. F., Hughes, T. R., Davierwala, A. P., Robinson, M. D., Stoughton, R., and Altschuler, S. J. (2002). Large-scale prediction of saccharomyces cerevisiae gene function using overlapping transcriptional clusters. *Nature Genetics*, 31:255–265.

[Wu and Manber, 1992] Wu, S. and Manber, U. (1992). Fast text searching allowing errors. *CACM*, 35(10):83–91.

[Wu et al., 1990] Wu, S., Manber, U., and Myers, G. (1990). An O(NP) sequence comparison algorithm. *Information Processing Letters*, 35:317–323.

[Yamanishi et al., 2000] Yamanishi, K., Takeichi, J., and Williams, G. (2000). On-line unsupervised outlier detection using finite mixtures with discounting learning algorithms. In *Proceedings of the 6th ACM SIGKDD International Conference on Knowledge Discovery and Data Mining*, pages 320–324.

[Yang et al., 2005] Yang, H., Burke, T., Dempsey, J., Diaz, B., Collins, E., Toth, J., Beckmann, R., and Ye, X. (2005). Mitotic requirement for aurora A kinase is bypassed in the absence of aurora B kinase. *FEBS Letters*, 579(16):3385–3391.

[Yang et al., 2003] Yang, J., Wang, H., Wang, W., and Yu, P. (2003). Enhanced biclustering on expression data. In *Proceedings of the 3rd IEEE Conference on Bioinformatics and Bioengineering*, pages 321–327.

[Yeang et al., 2001] Yeang, C.-H., Ramaswamy, S., Tamayo, P., Mukherjee, S., Rifkin, R. M., Angelo, M., Reich, M., Lander, E., Mesirov, J., and Golub, T. (2001). Molecular classification of multiple tumor types. *Bioinformatics*, 17(Supplement 1):316–322.

[Yeung et al., 2003] Yeung, K. Y., Medvedovic, M., and Bumgarner, R. E. (2003). Clustering gene-expression data with repeated measurements. *Genome Biology*, 4(5):R34.

[Yoon and Donahue, 1992] Yoon, H. and Donahue, T. F. (1992). Control of translation initiation in *saccharomyces cerevisiae*. *Molecular Microbiology*, 6(11):1413–1419.

[Yu and Yang, 2001] Yu, H. and Yang, J. (2001). A direct LDA algorithm for high-dimensional data — with application to face recognition. *Pattern Recognition*, 34(10):2067–2070.

[Yu et al., 2005] Yu, Y., Ji, H., Doudna, J. A., and Leary, J. A. (2005). Mass spectrometric analysis of the human 40s ribosomal subunit: native and HCV IRES-bound complexes. *Protein Science*, 14:1438–1446.

[Zelnik-Manor and Perona, 2004] Zelnik-Manor, L. and Perona, P. (2004). Self-tuning spectral clustering. In *Advances in Neural Information Processing Systems 17*, pages 1601–1608.

[Zhang et al., 2003] Zhang, H., Yu, C.-Y., and Singer, B. (2003). Cell and tumor classification using gene expression data: construction of forests. *Proceedings of the National Academy of Sciences of USA*, 100(7):4168–4172.

[Zhang et al., 1996] Zhang, T., Ramakrishnan, R., and Livny, M. (1996). BIRCH: An efficient data clustering method for very large databases. In Jagdish, H. V. and Mumick, I. S., editors, *Proceedings of ACM SIGMOD International Conference on Management of Data*, pages 103–114.

[Zhao et al., 2005] Zhao, H., Kim, Y., Wang, P., Lapointe, J., Tibshirani, R., Pollack, J. R., and Brooks, J. D. (2005). Genome-wide characterization of gene expression variations and DNA copy number changes in prostate cancer cell lines. *The Prostate*, 63(2):187–197.

[Zhou et al., 2006] Zhou, S., Herschleb, J., and Schwartz, D. C. (2006). A single molecule system for whole genome analysis. In Mitchelson, K. R., editor, *New Methods for DNA Sequencing*. Elsevier.

[Zhou et al., 2002] Zhou, X., Kao, M. C., and Wong, W. H. (2002). Transitive functional annotation by shortest-path analysis of gene expression data. *Proc. Natl. Acad. Sci. USA*, 99:12783–12788.

[Zhou et al., 2005] Zhou, X. J., Kao, M.-C. J., Huang, H., Wong, A., Nunez-Iglesias, J., Primig, M., Aparicio, O. M., Finch, C. E., Morgan, T. E., and Wong, W. H. (2005). Functional annotation and network reconstruction through cross-platform integration of microarray data. *Nature Biotechnology*, 23:238–243.

[Zhu et al., 2005] Zhu, Z., Shendure, J., and Church, G. M. (2005). Discovering functional transcription-factor combinations in the human cell cycle. *Genome Research*, 15:848–855.

[Zien et al., 2000] Zien, A., Rätsch, G., Mika, S., Schölkopf, B., Lengauer, T., and Müller, K.-R. (2000). Engineering support vector machine kernels that recognize translation initiation sites. *Bioinformatics*, 16(9):799–807.

[Ziv and Lempel, 1977] Ziv, J. and Lempel, A. (1977). A universal algorithm for sequential data compression. *IEEE Transactions on Information Theory*, 23(3):337–343.